LANDSCAPING

Principles & Practices

EIGHTH EDITION

Lab Manual

LANDSCAPING

Principles & Practices

EIGHTH EDITION

Lab Manual

Alissa F. Smith

Australia • Brazil • Mexico • Singapore • United Kingdom • United States

Landscaping: Principles and Practices, 8th Edition, Lab Manual

Alissa F. Smith

SVP, GM Skills & Global Product Management: Jonathan Lau

Product Director: Matt Seeley

Product Manager: Nicole Robinson

Executive Director, Content Design: Marah Bellegarde

Learning Design Director: Juliet Steiner

Learning Designer: Jennifer Starr

Product Assistant: Nicholas Scaglione

Vice President, Marketing Services: Jennifer Ann Baker

Marketing Manager: Abigail Hess

Senior Director, Content Delivery: Wendy Troeger

Associate Content Manager: Susan VandenBergh

Design Director: Jack Pendleton

Cover image credits: Garden: © 2009fotofriends/ Shutterstock.com; Retaining wall and patio: © iStockphoto/herreid; Beautiful spring flowers: © iStockphoto/elenaleonova; Natural landscaping: © Stok Görseller/Shutterstock.com; Limestone: © Kues/Shutterstock.com

For product information and technology assistance, contact us at **Cengage Customer & Sales Support, 1-800-354-9706**

For permission to use material from this text or product, submit all requests online at **www.cengage.com/permissions.** Further permissions questions can be e-mailed to **permissionrequest@cengage.com**

Library of Congress Control Number: 2017953223

ISBN: 978-1-337-40343-6

Cengage
20 Channel Center Street
Boston, MA 02210
USA

Cengage is a leading provider of customized learning solutions with employees residing in nearly 40 different countries and sales in more than 125 countries around the world. Find your local representative at **www.cengage.com.**

Cengage products are represented in Canada by Nelson Education, Ltd.

To learn more about Cengage, visit **www.cengage.com**

Purchase any of our products at your local college store or at our preferred online store **www.cengage.com**

Notice to the Reader

Publisher does not warrant or guarantee any of the products described herein or perform any independent analysis in connection with any of the product information contained herein. Publisher does not assume, and expressly disclaims, any obligation to obtain and include information other than that provided to it by the manufacturer. The reader is expressly warned to consider and adopt all safety precautions that might be indicated by the activities described herein and to avoid all potential hazards. By following the instructions contained herein, the reader willingly assumes all risks in connection with such instructions. The publisher makes no representations or warranties of any kind, including but not limited to, the warranties of fitness for particular purpose or merchantability, nor are any such representations implied with respect to the material set forth herein, and the publisher takes no responsibility with respect to such material. The publisher shall not be liable for any special, consequential, or exemplary damages resulting, in whole or part, from the readers' use of, or reliance upon, this material.

Printed at CLDPC, USA, 09-24

CONTENTS

SECTION TWO   LANDSCAPE CALCULATIONS

SECTION THREE   CUSTOMER SERVICE

SECTION FOUR   LANDSCAPE MAINTENANCE

PREFACE

Landscaping: Principles and Practices, 8th Edition, Lab Manual includes exercises that encourage students to apply what they have learned through their landscaping course readings or lessons. Various activities require students to further explore design principles, landscape plan development, and soft skills, to ensure that they are well-versed in these different aspects of the landscaping industry.

Most residential landscapes are planned, installed, and maintained by skilled workers with high school, technical school, or college training in landscape horticulture. As such, high school programs should include skills development in landscape design as well as installation and maintenance. This type of instruction will allow students to leave high school with highly sought-after employability skills.

The purpose of this lab manual is to provide a step-by-step guide to basic skills development in landscape design, installation, maintenance, and customer service. The materials can be adapted to large-group instruction or to the individualized format presented. In either case, the student should master one exercise before moving to another.

PRECISION EXAMS Precision Exam Alignment

This eighth edition of *Landscaping: Principles and Practices,* including text and accompanying lab exercises, is aligned to Precision Exams' **Landscape Management** exam, part of the **Agriculture, Food and Natural Resources** Career Cluster. The **Agriculture, Food and Natural Resources** pathway connects industry with skills taught in the classroom to help students successfully transition from high school to college and/or career. Working together, Precision Exams and National Geographic Learning/Cengage focus on preparing students for the workforce, with exams and content that is kept up to date and relevant to today's jobs. To access a corresponding correlation guide, visit the accompanying Instructor Companion Website for *Landscaping: Principles and Practices, 8th Edition.* For more information on how to administer the **Landscape Management** exam or any of the 170+ exams available to your students, contact your local NGL/Cengage Sales Consultant.

ABOUT THE AUTHOR

Alissa F. Smith is the Associate Executive Director of the National Association of Agricultural Educators (NAAE), which is the national professional organization for school-based agricultural educators across the nation. She is primarily responsible for national professional development for agricultural educators. Prior to working at NAAE, Alissa was a classroom agriculture teacher in Central Kentucky who specialized in teaching horticulture and agriscience. She developed curriculum and instructional material for high school agriculture classes that included Landscaping/Landscape Design, Greenhouse Technology, and Floriculture/Floral Design. Alissa also focused her time developing agriscience curriculum and serving on state committees to develop a state agriscience pathway in Kentucky.

Alissa completed her undergraduate degree in Agriculture at the University of Kentucky and holds a teaching certification in agriculture for grades 5 through 12. She also earned her master's degree in Agriculture from the University of Kentucky.

SECTION ONE

Landscape Design

Identifying Materials and Equipment

OBJECTIVE

To familiarize students with the tools of landscape drafting and the use(s) of each.

SKILLS

After studying this unit, you should be able to:

- Identify landscape drafting tools, by name, from line drawings or the actual tools.
- Select the tool needed for a particular task or function.

MATERIALS NEEDED*

Drafting board (minimum size 17" × 22")

T-square

Triangles (45°–45°–90° and 30°–60°–90°)

Adjustable triangle

2H drawing pencil

Lead holder

Vinyl eraser

Eraser shield

Engineer's scale (triangular)

Architect's scale (triangular)

French curves

Compass

Protractor

Circle template ($\frac{1}{16}$" to 1" circles, minimum)

Drafting paper (plain or gridded)

Pictures may be substituted for items not available.

INTRODUCTION

The landscape designer, like the architect or engineer, must be familiar with the use of basic drafting tools. Literally thousands of products are on the market today. However, the following is a description of basic tools with which the beginning student of landscape design should be familiar.

Drafting Boards

The drafting boards give a smooth surface for drafting on paper. It can be used on a table, desk, or other steady surface. Drafting boards come in various sizes and are made of smooth-sanded laminated wood or plastics. Wooden boards or tables should be covered with a vinyl drafting board cover. Such covers are more durable than wood and less subject to damage (Figure 1–1).

T-square

A T-square is used for drawing straight lines that are parallel to the edge of the drafting board. It may be used for either vertical or horizontal lines. A T-square is not required where boards or tables are equipped with a parallel bar or drafting machine (Figures 1–1 and 1–2).

Drafting Machines

Drafting machines replace T-squares, parallel bars, and triangles, and allow for more rapid drafting. A drafting machine is not essential for beginning students, but students in advanced classes, draftspeople, or professionals will find it the single most time-saving mechanical device available (Figure 1–3).

Drafting Paper

Drafting paper is available as opaque or transparent in a wide variety of sizes. Opaque papers are fine for rough drafts or final copies where the drawings are not to be blueprinted. When blueprints are desirable, it is essential to draw on high-quality vellum made of 100% cotton and labeled as 100% rag. For ease in drafting, paper with nonreproductive blue grids is available as opaque or transparent. When blueprinted or photocopied, the blue lines do not reproduce. Gridded paper is usually available in 4 × 4 ($\frac{1}{4}$" = 1'), 8 × 8 ($\frac{1}{8}$" − 1') or 10 × 10 ($\frac{1}{10}$" = 1') (Figure 1–4).

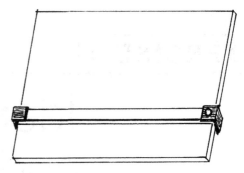

FIGURE 1–1 Drafting board and parallel bar.

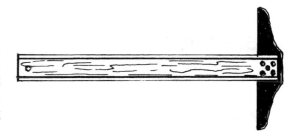

FIGURE 1–2 T-square.

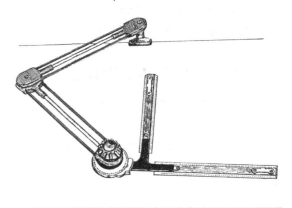

FIGURE 1–3 Drafting machine.

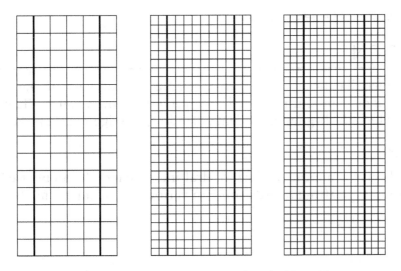

FIGURE 1–4 Gridded papers—4 × 4, 8 × 8, 10 × 10.

Triangles

Triangles are used to draw angled lines off any straight line. The most commonly used angles are 45 degrees and 90 degrees. Angles other than 30, 45, 60, or 90 degrees can be drawn with the aid of a protractor or an adjustable triangle (Figure 1–5).

Adjustable Triangles

These triangles cost considerably more than standard fixed triangles but can be used to measure or plot any angle between 0 and 90 degrees in 1-degree increments (Figure 1–6).

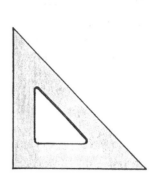

FIGURE 1–5 Triangles.

Protractor

A protractor is used to measure the angle of any two joining lines, from 0 to 180 degrees, in 1-degree increments for the standard protractor or 0 to 360 degrees for the round version. It is used to determine existing angles or to mark nonstandard angles not available on triangles (Figure 1–7).

Drawing Pencils

Drawing pencils come in varying degrees of hardness, usually 2B, B, HB, F, H, and 2H through 9H. Drafting is done with lead having an HB rating or higher. An H-rated pencil contains harder lead and produces lighter lines that are less likely to smear. Avoid leads with a B rating because B leads are much softer and are most often used for artwork or sketching. A good choice for the beginning landscape designers is in the range of HB to 2H (Figure 1–8).

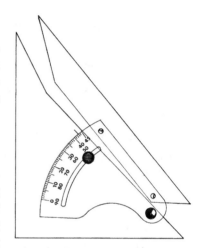

FIGURE 1–6 Adjustable triangle.

Lead Holders

Lead holders, or mechanical pencils, are more expensive initially but are much less expensive over the long run. The standard lead holder uses lead 2 millimeters (mm) thick and requires a lead pointer. Some prefer lead holders with smaller lead, which are available in 0.3-, 0.5-, 0.7-, or 0.9-millimeter thicknesses. These holders do not usually require sharpening of the lead, but because of the smaller lead, breakage of the point is more frequent (Figure 1–8).

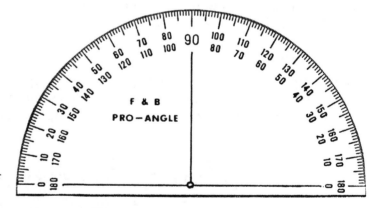

FIGURE 1–7 Protractor.

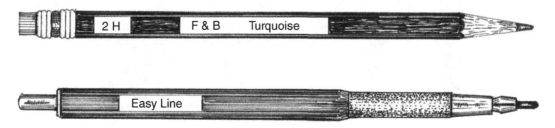

FIGURE 1–8 2H pencil and lead holder.

Pencil Sharpeners and Lead Pointers

Drafting pencils may be sharpened with any quality standard sharpener, either mechanical or electric. Lead pointers are essential for lead holders using 2-millimeter leads. Most pointers come with replaceable carbide blades, and some allow for adjustment of point taper (Figure 1–9).

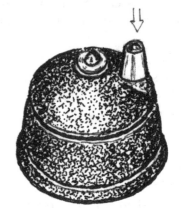

FIGURE 1–9 Two popular styles of lead pointers.

Erasers

Erasers should completely erase errors or changes without rubbing holes in the drafting paper. The best drafting erasers are made of vinyl and are nonabrasive (Figure 1–10). Electric erasers are faster but are not essential for the beginning student of design.

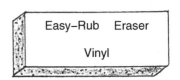

FIGURE 1–10 Vinyl eraser.

Eraser Shields

Eraser shields are used to erase unwanted lines or parts of lines without smudging or erasing desired lines. Shields have various sizes and shapes of openings to allow you to match the line to be erased. Eraser shields are inexpensive, and every draftsperson should have at least one (Figure 1–11).

Circle Templates

Circle templates allow rapid drawing of circles to represent trees, shrubs, and other landscape features. Templates chosen should have a variety of circle sizes (Figure 1–12). Templates appropri-

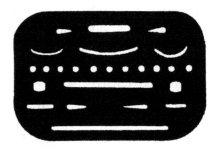

FIGURE 1–11 Eraser shield.

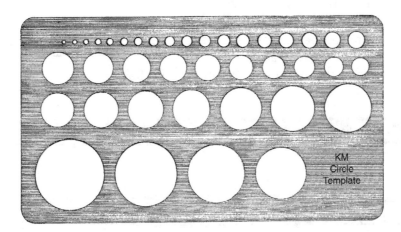

FIGURE 1–12 Circle template.

ate for landscape should have circles from $^1/_{16}$ to 1 inch minimum. A template with $^1/_{16}$- to $2^1/_4$-inch circles is more useful. Extra-large circles can be drawn with a compass.

Compass

A compass or bow compass is an adjustable instrument used to draw circles or arcs. It has two legs—one containing a metal point and the other containing a pencil or lead. The distance between the two legs is one-half the diameter of the circle. For example, a 1-inch setting will yield a 2-inch circle. Select a compass with a threaded adjustment bar to prevent slippage and maintain adjustment while drawing (Figure 1–13).

Scales

Scales are used for measuring or drawing lines that represent smaller dimensions on paper than they are actually in "real life." For example, a 1:10 scale means that 1 inch on paper represents 10 feet of real space. Triangular scales are popular because each has six or more scales on one instrument. Engineer's scales contain scales of $^1/_{10}$, $^1/_{20}$, $^1/_{30}$, $^1/_{40}$, $^1/_{50}$, and $^1/_{60}$.

Architect's scales usually contain scales of $^1/_2$, $^1/_4$, $^1/_8$, $^1/_{16}$, $^3/_8$, $^3/_4$, and $^3/_{16}$. Landscape designers commonly use scales of $^1/_{10}$ or $^1/_8$ (Figure 1–14).

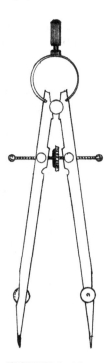

FIGURE 1–13
Bow compass.

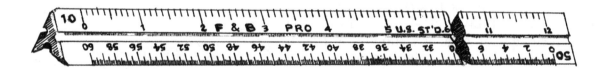

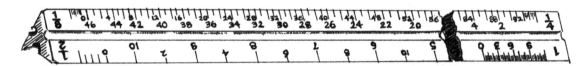

FIGURE 1–14 Engineer's scale (top) and architect's scale (bottom).

Lettering Guides

Lettering guides are used to draw guidelines essential to attractive lettering. The Ames® Lettering Guide is an inexpensive, adjustable guide with many different spacings, both standard and metric (Figure 1–15).

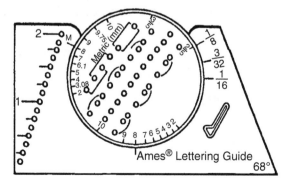

FIGURE 1–15 Ames® Lettering Guide.

Nonadjustable guides usually have only four line widths, but they allow for rapid drafting of guidelines (Figure 1–16).

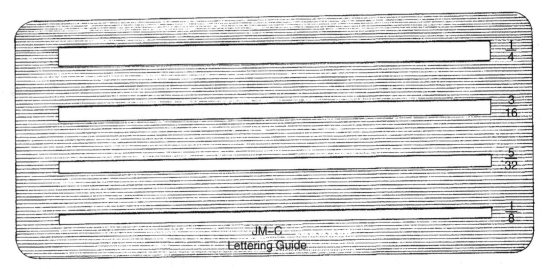

FIGURE 1–16 Nonadjustable lettering guide.

French Curves

French curves serve as a guide in making professional-looking curved lines for bed areas, walks, drives, and so on. The degree of curve can be changed by moving the french curve to an edge or section of the sketch having the same degree curve. There are many curves available in many price ranges. A package of three or four is usually adequate for landscape design (Figure 1–17).

FIGURE 1–17 French curves.

PUT IT INTO PRACTICE

1. Review the pictures included in this exercise, or examine the actual instruments (if provided by your instructor).

2. Complete the questions without the assistance of the reading in this exercise.

Exercise 1 Activity

Student Name _____ Date _____ Score _____

Fill in the Blanks

Fill in the blanks with the best answers.

1. A 45-degree line can be drawn quickly with the aid of a _____.

2. _____ _____, _____ _____, and _____ are useful in drawing lines parallel to the edge of the drafting board.

3. A(n) _____ _____ or a(n) _____ is useful in drawing nonstandard angles such as 37 degrees.

4. A _____ _____ allows rapid drawing of circles to indicate trees and shrubs.

5. An H rating indicates that a pencil has _____ lead.

6. Do not use leads with a _____ rating when drafting.

7. _____ _____ give a smooth, professional look to curving lines.

8. The best drafting erasers are made of _____.

9. Very large circles may be drawn with a _____.

10. _____ _____ allow you to erase a line without erasing adjacent or adjoining lines.

11. An _____ scale contains scales that are multiples of 10.

12. A 1:20 scale means that 1 inch equals _____.

Identification

Please identify the following items in their corresponding blank spaces:

1. Item A _____   5. Item E _____

2. Item B _____   6. Item F _____

3. Item C _____   7. Item G _____

4. Item D _____   8. Item H _____

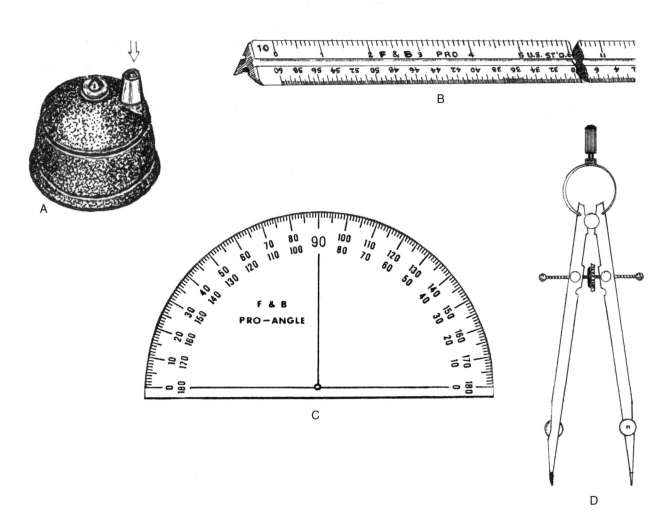

A

B

C

D

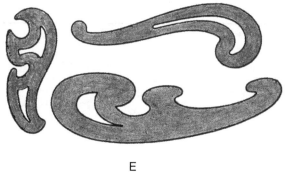

E

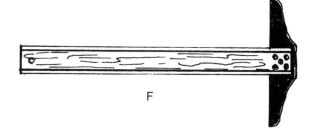

F

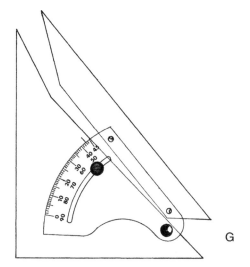

G

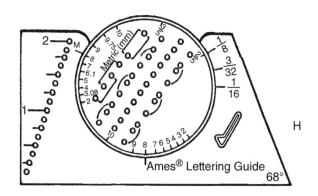

H

NOTES

Using Equipment and Working with Scale

OBJECTIVE

To provide hands-on practice in the use of equipment and materials for landscape drafting.

SKILLS

After studying this unit, you should be able to:

- Demonstrate a knowledge of scale by completing a basic scale drawing.
- Convert a drawing on a 1:20 scale to 1:10, or convert a drawing on a 1:16 scale to 1:8.
- Use various drafting instruments to complete the activity.

MATERIALS NEEDED

Drawing board

T-square

Triangles

Drawing pencil (HB, F, H, or 2H)

Vinyl eraser

Eraser shield

Engineer's scale or architect's scale

French curves

Compass

Protractor

Drafting paper (with or without grids) $8^1/_2 \times 11$–inch standard vellum

Drafting tape (not masking tape!)

Note: This exercise can be completed without a T-square and drawing board if paper with a fade-out grid is used.

INTRODUCTION

A landscape plan must be accurate—just like the design of a building or any other design that uses space. In order to be accurate, it must be drawn to a chosen scale. Simply stated, scale is a miniature representation of the real thing. A landscape plan should indicate the exact number of plants needed while giving a visual reference of the space needed in comparison to lawn areas or artificial features.

Before you can draw a landscape plan that will be accurate and usable, it is essential to have experience with the correct use of appropriate tools. As you complete this exercise, it is essential that you be perfectly honest. If you feel you do not understand the proper use of a tool, ask your instructor for additional help or practice.

As you work with drafting tools, you will come to realize that scale is the basis of all such tools. You must understand scale first and foremost. All other tools exist to help you draft plans to scale with ease and accuracy.

The word *scale* is also used to describe the measuring instrument used in drawing. On the design level, you should refrain from using the word *ruler* in favor of the more descriptive word *scale*. Two such scales are used in landscape design: the engineer's scale (Figure 2–1) and the architect's scale (Figure 2–2). Such scales are usually triangular, allowing for 6 different scales on the engineer's instrument and 11 on the architect's.

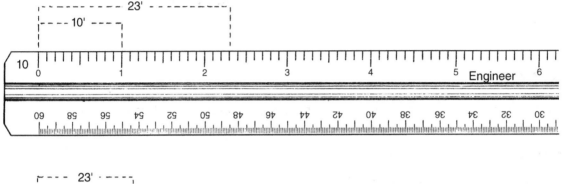

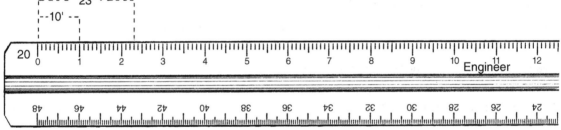

FIGURE 2–1 Engineer's scale.

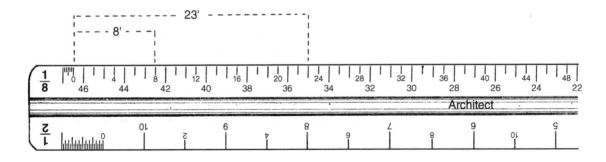

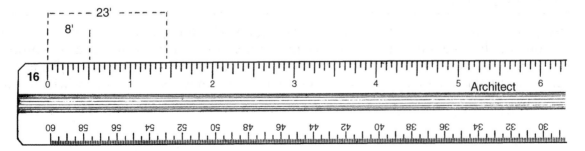

FIGURE 2–2 Architect's scale.

The engineer's scale is popular with landscape designers. For most residential properties, a scale of 1:10 will allow the property and building(s) to be placed on a standard sheet of vellum drafting paper (17" × 22", 24" × 36", 30" × 42", or 36" × 48"). On larger residential properties, a scale of 1:20 may be used; however, some symbols become very small at this scale. A better choice would be to divide the property into two or more areas and use a 1:10 scale to draft the plan on two or more sheets of vellum.

A 1:10 scale means that 1 inch is equal to 10 feet. Another way of looking at this is to say that $^1/_{10}$ inch is equal to 1 feet. In either case, every mark on the scale represents 1 foot. On the 1:20 scale 1 inch is equal to 20 feet. A drawing on this scale is exactly one-half the size of the same drawing on a 1:10 scale (Figure 2–1).

You will note that for every 10 feet, a longer mark exists on the scale, and a number is present. Just add 0 (zero) to the number on the scale. For example, on the 1:20 scale, the number 1 is actually $^1/_2$ inch and represents 10 feet. The number 2 is actually 1 inch and represents 20 feet.

Some designers prefer the architect's scale. The most popular architect's scale for landscape designers is the 1:8, where 1 inch = 8 feet, or $^1/_8$ inch = 1 foot. For larger properties, the 1:16 scale can be used, and on tiny properties a 1:4 scale is often appropriate.

As you study the architect's scale, you will note that the $^1/_8$-inch scale and the $^1/_4$-inch scale are located on the same edge. The $^1/_8$-inch scale reads left to right, whereas the $^1/_4$-inch scale reads right to left. This makes the $^1/_8$-inch scale somewhat more confusing at first, but with practice you will learn to use one while ignoring the other (Figure 2–2).

Everything you draw, except for "thumbnail sketches" in Exercise 13, should be drawn to scale. This is indicated on the drawing as "Scale: 1" = 10', Scale: 1" = 4'," and so on. This is usually noted just outside the property line on a completed drawing, as shown in Figure 12–1 (Exercise 12) or near the bottom of the drawing.

In addition to writing the scale being used, a useful symbol is the scale indicator bar (Figure 2–3). A scale indicator bar is simple to draft and should be included on any final drawings, as shown in Figure 12–1. With a minimum of practice, you will be able to draw the bar very quickly. The purpose of the scale indicator bar is to show the scale as a length of space. In copying a drawing, scale is sometimes slightly altered. In addition, you may desire to enlarge or reduce the drawing, and the scale bar can be used to determine the size of a feature, because everything will be reduced or enlarged proportionately.

Study Figures 2–1, 2–2, and 2–3. Be sure that you understand the scale you will use before you begin the activity.

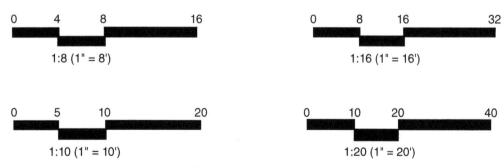

FIGURE 2–3 Sample scale indicator bars.

PUT IT INTO PRACTICE

This activity will enable you to practice scale and use some of the tools you studied in Exercise 1. Please have those tools readily available. You will also need two $8^{1}/_{2} \times 11$–inch sheets of drafting paper (standard vellum is fine if you have a drawing board and T-square). Tape all four corners to your drawing surface with the 11-inch side horizontal. If you are using a 10×10 or an 8×8 grid and a T-square, you must align the grid lines with the T-square.

1. Use the drawings that follow to complete this exercise. Once completed and evaluated, these drawings will continue to be used as we develop a complete landscape plan.

    a. The drawing in Activity I is drawn on a scale of 1:20. Draw the same residence on a 1:10 scale on your own drafting paper. Write the scale you are using, and draw a scale indicator bar on the same line. Your drawing will be much larger—actually twice as large. Please also review the Evaluation rubric that is included to know how your drawing will be assessed.

    b. The drawing in Activity II is drawn on a scale of 1:16. Draw the same residence on a 1:8 scale on your own drafting paper. Write the scale you are using, and draw a scale indicator bar on the same line. Please also review the Evaluation rubric that is included to know how your drawing will be assessed.

    NOTE: Measure each line with the scale indicated, either 1:20 or 1:16, and then draw each line using the requested scale, 1:10 or 1:8, as indicated in each activity above.

NOTES

Exercise 2 Activity I

Student Name _____ Date _____ Score _____

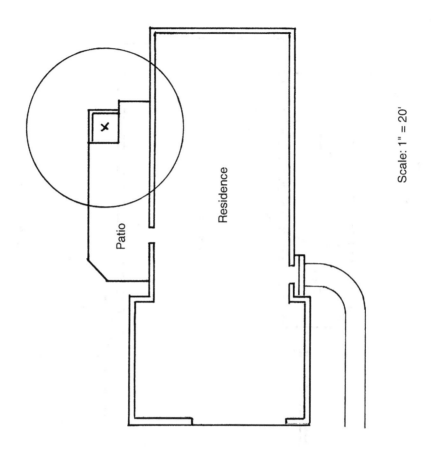

Evaluation

Consideration	Points	Student Score	Instructor Score
Exact scale achieved	70		
Lines are smooth and consistent	10		
Scale indicator bar is accurate	10		
Drawing is neat	10		
Total	100		

Exercise 2 Activity II

Student Name _____ Date _____ Score _____

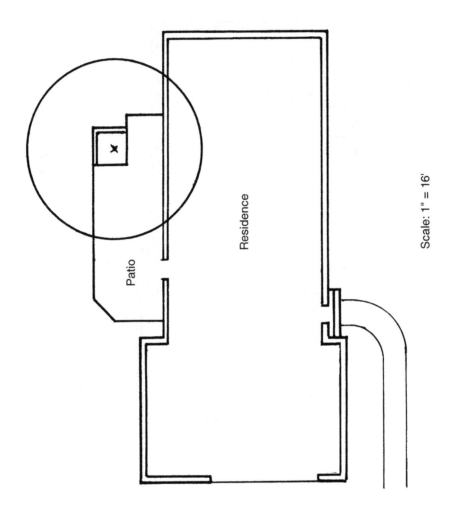

Patio

Residence

Scale: 1" = 16'

Evaluation

Consideration	Points	Student Score	Instructor Score
Exact scale achieved	70		
Lines are smooth and consistent	10		
Scale indicator bar is accurate	10		
Drawing is neat	10		
Total	100		

Developing Lettering Skills

OBJECTIVE

To develop good lettering skills through supervised practice.

SKILLS

After studying this unit, you should be able to:

- Properly use guidelines when lettering.
- Draw letters of consistent height and spacing.
- Draw letters using a single stroke.
- Demonstrate your lettering style.

MATERIALS NEEDED

Drawing board and T-square

Engineer's scale or architect's scale

Eraser and shield

Drawing pencil (HB, F, H, or 2H)

One $8^1/_2 \times 11$–inch sheet drawing paper

Drafting tape

House plan drawn in Exercise 2

INTRODUCTION

Good lettering is essential to give a professional or distinctive look to a drawing. It can "sell" both you and your plan to a prospective client. On the other hand, poor lettering is noticeable on a drawing, and it can cause others to doubt your ability as a designer. Time spent developing your lettering skills is time well spent.

Lettering is a developed skill. Although some people master lettering more quickly than others, everyone can develop acceptable lettering skills through practice. The following explanations should prove helpful:

1. Always use guidelines. Even the most experienced professionals use them. Horizontal guidelines are drawn with the aid of a T-square or, preferably, a lettering guide. To letter, you must have two guidelines, with the spacing between them determining the size of the letters. First, draw the top guideline. Using a scale, measure down the desired distance and place a dot. Then, using the dot as a reference, draw the second guide. A lettering guide eliminates this procedure. Always draw

guidelines lighter than your lettering, and never erase them after lettering. A blueprint or photocopy may pick up your guidelines, but that is acceptable so long as the lettering is darker.

Vertical guidelines are used for margins, columns, or as a guide for centering. Refer to Exercise 28 for usage of vertical guidelines.

2. All lettering should be parallel with the bottom of the sheet. This enables you to read everything on the sheet without having to turn it.

3. Letters should touch both the top and bottom guidelines. This gives even height to the letters.

4. Letters are usually drafted in capitals; however, lowercase—even cursive—letters have been used effectively by architects and designers. Caution should be exercised in attempting to get too fancy.

5. Don't go over a line twice. This makes that part of the letter darker, and it will stand out on the blueprint or photocopy. If you need to correct the letter, erase it fully and do it over.

6. Avoid wavy lines. Wavy lines are the result of marking too slowly. Relax and make deliberate strokes.

7. Avoid making letters too narrow or too wide. This will take practice but will soon become second nature (Figure 3–1).

NOTE: Use guidelines in your practice even if you are using fade-out vellum. Guidelines should become a habit with you, because there will be many times when you will have to letter on nongrid paper (Figure 3–2).

LETTERS SHOULD TOUCH BOTH GUIDELINES

AVOID UNEVEN HEIGHT

TOO NARROW TOO WIDE

DON'T GO OVER LETTERS TWICE

USE SINGLE STROKES

FIGURE 3–1 Do's and don'ts of lettering.

FIGURE 3–2 Sample lettering.

PUT IT INTO PRACTICE

Tape an $8^1/_2 \times 11$–inch sheet of vellum to your drawing surface and align the bottom or top edge using your T-square. Complete the following activity.

1. Draw two pairs of horizontal guidelines for several different size letters. Draw a vertical guideline for your left margin.

2. Draw two alphabets (including numbers) using $^2/_{10}$-inch letters for engineer's scales or $^3/_{16}$-inch for architect's scales. Draw the standard alphabet first, and then draw one of the slanting alphabets (Figure 3–3).

3. Draw two alphabets for $^1/_{10}$-inch letters or $^1/_8$-inch letters, using the style you prefer.

4. Write your own first and last name, using both lettering sizes.

5. Write the sentence, "Good lettering skills are acquired through regular practice" (Figure 3–3).

6. Using the house plan you drew for Exercise 2: Activity I, label the rooms and areas as shown in Figure 3–4. Save your house plan. It will be used for Exercise 4.

NOTES

ABCDEFGHIJKLMNOPQRST
UVWXYZ 1234567890 &
abcdefghijklmnopqrstuvw
xyz

ABCDEFGHIJKLMNOPQRSTUVWX
YZ   1234567890   &

ABCDEFGHIJKLMNOPQRSTUVWXYZ   12345
67890   abcdefghijklmnopqrstuvwxyz

ABCDEFGHIJKLMNOPQRST
UVWXYZ 1234567890 &
abcdefghijklmnopqrstuvwxyz

ABCDEFGHIJKLMNOPQRSTUVW
XYZ   1234567890   &

GOOD LETTERING SKILLS ARE ACQUIRED THROUGH
REGULAR PRACTICE.

FIGURE 3–3 Various lettering sizes.

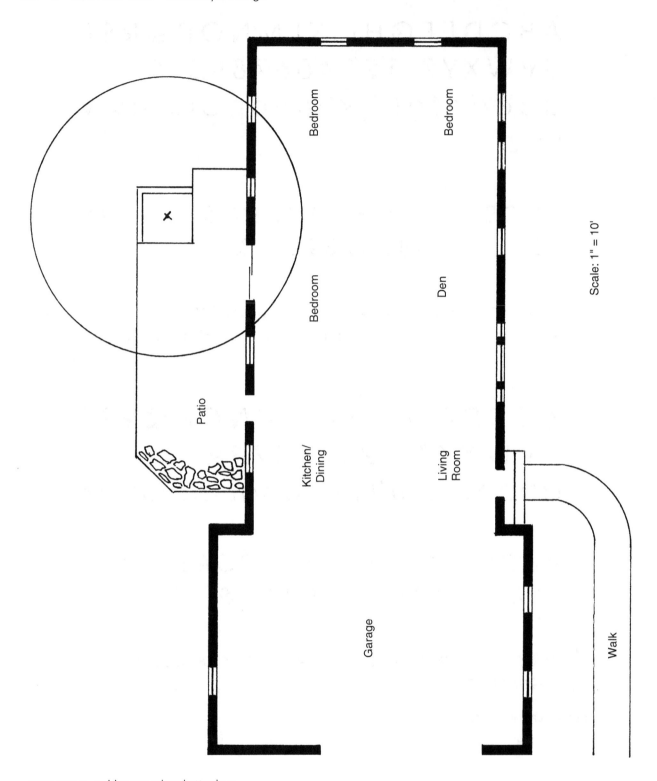

FIGURE 3–4 House plan lettering.

Scale: 1" = 10'

Bedroom

Bedroom

Bedroom

Den

Patio

Kitchen/
Dining

Living
Room

Garage

Walk

Exercise 3 Activity

Student Name _____ Date _____ Score _____

Evaluation

Consideration	Points	Student Score	Instructor Score
Letters touch guidelines	50		
Letters are equal width	10		
Spacing is acceptable	10		
Pencil strokes are even	10		
Overall neatness	20		
Total	100		

NOTES

Using Symbols in Drawing the Residence

OBJECTIVE

To gain experience in drawing and recognizing symbols for walls, windows, doors, and surfacing materials for a residence.

SKILLS

After studying this unit, you should be able to:

- Draft walls, windows, and doors of a residence.
- Draft surfacing materials for patios, walks, and drives of a residence.
- Darken walls of a residence plan with uniform shading.

MATERIALS NEEDED

House plan from Exercise 3

Drawing board and T-square

Engineer's scale or architect's scale

Eraser and shield

Drawing pencil (HB, F, H, or 2H)

INTRODUCTION

In drafting a residence, it is essential that the outside walls, windows, and doors be located accurately. This will determine the kind, number, size, and placement of the foundation plants (plants near the wall or foundation of a structure). To be usable, a landscape plan must be an accurate representation of the actual house and property.

It is not necessary to draw inside walls when drafting the residence; however, it is useful to indicate the location of rooms on the drawing, as you did in Exercise 3. This is helpful to the designer in determining the views from a particular room or in locating other features in the landscape. For example, a satisfying or interesting view from the den or family room might be a primary consideration in developing the total garden design.

Once the general dimensions of a residence are drawn, the windows and doors should be the next consideration. If you are working from an existing architect's drawing, you can simply determine the size and location of features and draw them on your plan.

Start by measuring from a corner to the first window or door of the wall. Always check your measurement from both corners of a wall, and recheck often as you draw. Use the same procedure if you are taking measurements of the actual residence, using a tape measure or other device.

The exact sizes of windows or doors vary greatly. For instance, a window might be 28 inches, 32 inches, or some other size. Because your engineer's scale does not show inches, you will have to estimate small measurements (Figure 4–1). A good rule of thumb for the landscape designer is to round off such measurements to the nearest one-half foot, or 6 inches. It is best to round off to the next largest one-half foot. This will help ensure that plants that are proposed near windows will not cover or obstruct windows. The same procedure of rounding off to the nearest one-half foot can be used in drafting doors.

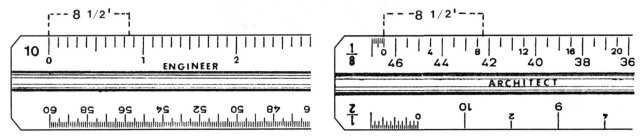

FIGURE 4–1 Estimating measurements.

Standard windows can usually be drawn at $2^1/_2$ to 3 feet. Bathroom windows are often smaller and measure 2 to $2^1/_2$ feet. Picture windows and bay windows can be almost any size, but they usually vary from 6 to 8 feet in length. Doors usually vary between $2^1/_2$ and $3^1/_2$ feet.

Once windows and doors are located, the next step is to darken the remaining wall space. This serves to make the residence stand out on the plan, and it makes windows and doors stand out. Strive to use the same degree or density (darkness) in shading the walls (Figure 4–2).

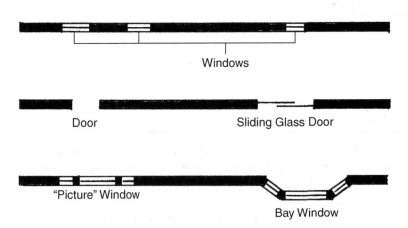

Windows

Door   Sliding Glass Door

"Picture" Window

Bay Window

FIGURE 4–2 Locating windows and doors and darkening wall space.

Patios, drives, walks, and paths are made of concrete, stone, or brick. (Wooden decks will be covered in another exercise.) It is not necessary to draw symbols for the entire area. Drawing symbols to show the material for a section of the area is sufficient. Other surfacing materials, such as asphalt, can simply be labeled as such on the plan (Figure 4–3).

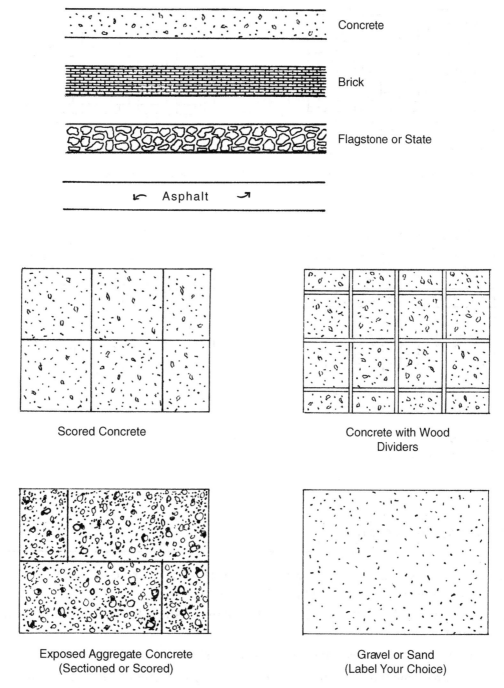

Concrete

Brick

Flagstone or State

↳ Asphalt ↗

Scored Concrete

Concrete with Wood Dividers

Exposed Aggregate Concrete (Sectioned or Scored)

Gravel or Sand (Label Your Choice)

FIGURE 4–3 Various materials used for patios, drives, walks, and paths. *(continued on next page)*

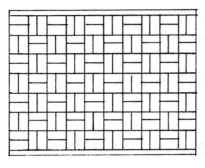

Brick—Basketweave Pattern

Brick—Herringbone Pattern

Brick—Running Bond Pattern
(With Crosstie Dividers)

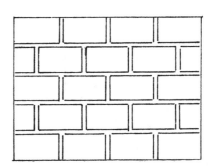

Cut Stone or Concrete
(With Mortar Joints)

FIGURE 4–3 Various materials used for patios, drives, walks, and paths. *(continued from previous page)*

PUT IT INTO PRACTICE

1. Using the house plan drawn in Exercise 2 and lettered in Exercise 3, draw the windows in the same location as shown in Figure 3–4 of Exercise 3.

   a. Erase part of the wall connected to the patio to allow for the sliding glass door; then draft it as shown.

   b. Darken the wall spaces between windows, as shown.

2. Using appropriate symbols, indicate the material you desire for the front walk and rear patio. You may select the concrete and flagstone drawn on the plan in Figure 3–4, or you may select and draw another material. Use a 1:10 scale.

Exercise 4 Activity

Student Name _____ Date _____ Score _____

Evaluation

Consideration	Points	Student Score	Instructor Score
Walls are darkened and accurate	10		
Windows are located accurately	30		
Doors are located accurately	30		
Surfacing symbols are accurate and acceptable	10		
Overall neatness	20		
Total	100		

NOTES

Using Symbols in Drawing Trees

OBJECTIVE

To provide experience in drawing and recognizing symbols of trees on a landscape plan.

SKILLS

After studying this unit, you should be able to:

- Draw symbols to represent small, medium, and large trees.
- Draw symbols for both evergreen and deciduous trees.
- Create symbols of your own.

MATERIALS NEEDED

Drawing board and T-square

Drawing pencil (HB, F, H, or 2H)

Eraser and shield

Engineer's scale or architect's scale

Circle template ($^1/_{16}$" to $2^1/_4$")

One $8^1/_2 \times 11$–inch sheet of drafting paper

Drafting tape

INTRODUCTION

In designing landscapes, most permanent plants are represented by circular symbols. On smaller or partial plans containing a relatively small number of plants, it is appropriate to represent plants by circles drawn with a circle template. However, on larger plans with a greater number of plants, individual plants become lost in an endless "jungle" of simple circles.

The purpose of symbols is to represent variety in plants and, if desirable, to indicate whether the plant is deciduous (loses its leaves in fall) or evergreen (has foliage all year). Whereas perfect circles are not necessary in symbolizing plants, it is important to keep them mostly round.

Large trees are those trees that mature at 40 feet or taller. Medium trees mature at 20 to 40 feet, and small trees mature at 10 to 20 feet.

The following guidelines will work for the majority of trees in your plan:

1. Large trees   20 feet in diameter or greater

2. Medium trees   15 to 20 feet in diameter

3. Small trees   10 to 15 feet in diameter

The diameter measurements indicate trunk diameter.

To draw a tree symbol, decide the diameter circle you need. Choose a circle that is the desired size from the circle template. It might be helpful to place your scale over the circle template and measure various circles. If you do not have a circle large enough on your template, a compass can be used to draw the circle.

Once a circle is selected, draw it lightly, as you did in drawing guidelines for lettering. Next, decide upon a symbol and begin to draw the plant. You will note from the sample provided with this exercise that evergreen symbols tend to have points or straight lines as part of the symbol. Deciduous symbols are less rigid or less pointed. Always mark the center of the plant, as this indicates the exact location of the initial planting (Figures 5–1 and 5–2).

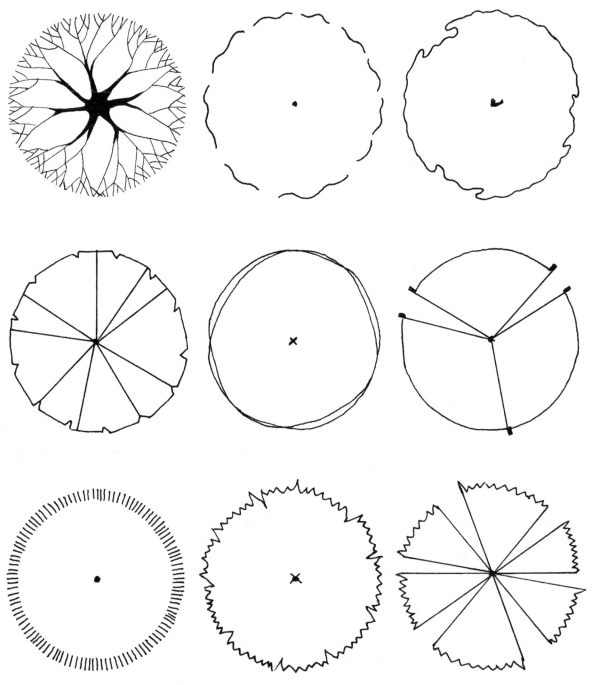

FIGURE 5-1 Large tree symbols.

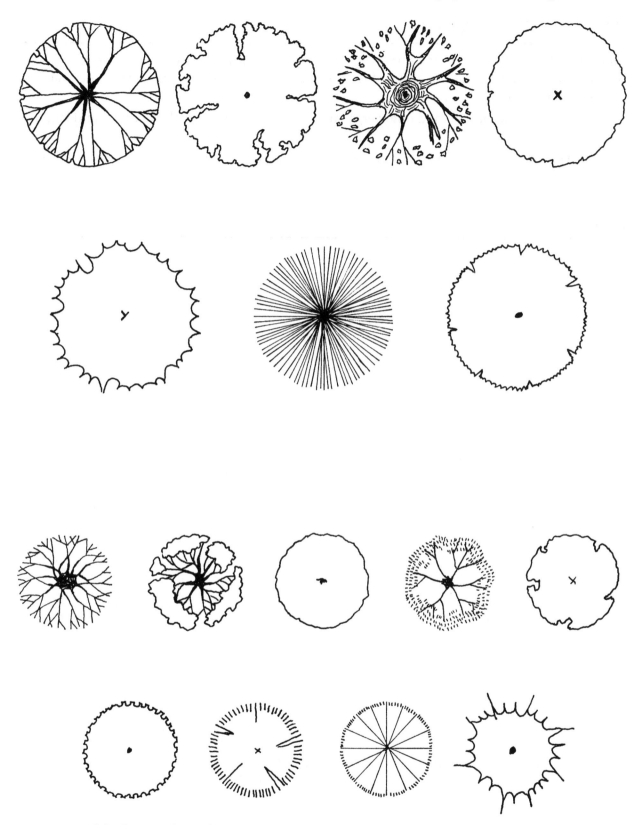

FIGURE 5–2 Medium and small tree symbols.

PUT IT INTO PRACTICE

Obtain an $8^{1}/_{2} \times 11$–inch sheet of drafting paper and tape it to your drawing surface. Draw the following in 1:10 or 1:8 scale. Be neat and strive for smooth, even lines. Please review the Evaluation rubric before beginning.

1. Draw two deciduous and one evergreen for each size group (small, medium, and large), for a total of nine trees.

2. Design your own symbols and draw one each of large, medium, and small trees, for a total of three additional trees.

NOTES

Exercise 5 Activity

Student Name _____ Date _____ Score _____

Evaluation

Consideration	Points	Student Score	Instructor Score
Measurements are accurate	50		
Symbols are uniform and circular	20		
Dots are centered	15		
Overall neatness	15		
Total	100		

NOTES

Using Symbols in Drawing Shrubs

OBJECTIVE

To provide experience in drawing and recognizing symbols for shrubs on a landscape plan.

SKILLS

After studying this unit, you should be able to:

- Draw symbols to represent large, medium, and dwarf shrubs.
- Draw groupings of shrubs.
- Create symbols of your own.

MATERIALS NEEDED

Drawing board and T-square

Drawing pencil (HB, F, H, or 2H)

Eraser and shield

Engineer's scale or architect's scale

Circle template ($^1/_{16}$" to $2^1/_4$")

One $8^1/_2 \times 11$–inch sheet of drafting paper

Drafting tape

INTRODUCTION

With the exception of lawn grasses, shrubs usually make up the largest total number of plants in a landscape. Most shrubs are present in groups of the same species of plants or as a single plant that is part of a group. Plants that are part of a group must be drawn accurately to prevent overcrowding in the actual landscape. This occurs when the symbols are drawn too small. On the other hand, if the symbols are drawn too large, the planting will appear too thin in real life. Therefore, one should draft the symbol to the size of a particular plant at average mature width. This requires a source of information on landscape plants, and there are many books that provide such information. In addition, many free or inexpensive materials are available from the United States Department of Agriculture (USDA) or your local extension service. Reputable nurseries are usually happy to share information.

Shrubs vary in both width and height, but height is usually the measurement used in placing plants in size groups. Dwarf shrubs are plants that grow to less than 4 feet in height at maturity.

Medium shrubs grow to 4 to 6 feet, and large shrubs are over 6 feet in height (Figure 6–1). Different publications offer slightly different numbers.

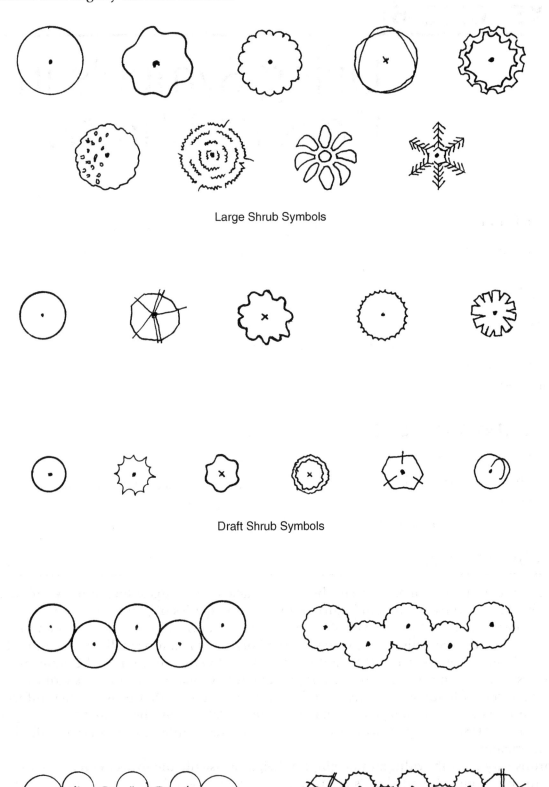

Large Shrub Symbols

Draft Shrub Symbols

Sample Groupings

FIGURE 6–1 Shrub symbols.

In the absence of specific information on width, the following guidelines work for most plans:

1. Dwarf shrubs   3 to 4 feet in diameter

2. Medium shrubs   5 to 6 feet in diameter

3. Large shrubs   6 to 9 feet in diameter

When drafting the shrubs in a group, allow the symbols to touch or appear joined, as illustrated in Figure 6-1, Sample Groupings. This will allow shrubs in the actual landscape to look "full," without being overcrowded.

PUT IT INTO PRACTICE

Tape a sheet of $8^{1}/_{2} \times 11$–inch drafting paper to your drawing surface. Use the 1:10 scale to complete the following activity. Review the Evaluation rubric before beginning this activity.

1. Draw three individual shrubs of each size group—dwarf, medium, and large—for a total of nine symbols.

2. Design your own symbols, one each for dwarf, medium, and large, for a total of three shrubs.

3. Draw a grouping of five shrubs for each size group.

NOTES

NOTES

Exercise 6 Activity

Student Name _____ Date _____ Score _____

Evaluation

Consideration	Points	Student Score	Instructor Score
Measurements are accurate	50		
Groupings are uniform	15		
Symbols are neat in appearance	20		
Dots are centered	15		
Total	100		

NOTES

Using Symbols in Drawing Ground Covers, Vines, and Flower Beds

OBJECTIVE

To provide experience in drawing and recognizing symbols for ground covers, vines, and flowers on the landscape plan.

SKILLS

After studying this unit, you should be able to:

- Draw symbols to represent both broadleaf ground covers and narrow leaf ground covers.
- Draw symbols to represent annual flowers in the landscape.
- Draw symbols to represent vines.
- Understand the major purpose(s) of ground covers, vines, and annual flowers in the landscape.

MATERIALS NEEDED

Drawing board and T-square

Drawing pencil (HB, F, H, or 2H)

Eraser and shield

Engineer's scale or architect's scale

French curve

45°–45°–90° triangle

Drafting tape

INTRODUCTION

Ground covers are permanent, low-growing plants that take the place of turf (grass) in a landscape. Almost every garden has an area that is suitable for some type of ground cover. Ground cover is easier to maintain than turf and has increased in popularity. Ground covers are available in a wide variety of forms, leaf shapes and sizes, flowers, color, and other characteristics. Ground covers are best used in combination with other landscape plants, especially trees. Steep embankments and shady areas under low-branched trees are ideal areas for ground covers.

In drawing ground covers, we do not show individual plants because many are tiny compared to other plants. We show ground covers as a mass of plants, and we illustrate the shapes and boundaries of the bed areas. In using ground covers, a definite area should be established and maintained because many of the plants spread by either roots or stems. Most can be contained by monthly maintenance shearing during the warm season. Figure 7–1 shows possible symbols for drawing ground covers.

Vines are used in the landscape in many ways. As with all plant groups, much variety exists. Vines are "climbing" plants that may or may not require support or attachment. Vines may be grown against a bare wall; on a fence, pole, or arbor (overhead beams); or on a trellis (upright latticework or crossed slatting). Carefully placed vines can add much interest to the landscape.

Herbaceous flowers are used as accent plants. A bed planted in flowers will provide interest and color throughout warm weather. In designing areas for annual flowers in the landscape, it is not necessary to name every plant, because annual flowers may be changed from year to year or season to season. If perennial flowers are desired, more care in selection is required, because they emerge from the roots each spring. Flower beds should be planned in such a way that the absence of flowers will not detract from the landscape when they are not present. Symbols may or may not be necessary, but the areas should be labeled *Flowers* (Figure 7–1).

PUT IT INTO PRACTICE

1. Follow directions, and complete the Exercise 7 Activity. Draw directly on the sheet provided. Use a 1:10 or a 1:8 scale. Please review the Evaluation rubric before beginning.

NOTES

Ground Cover

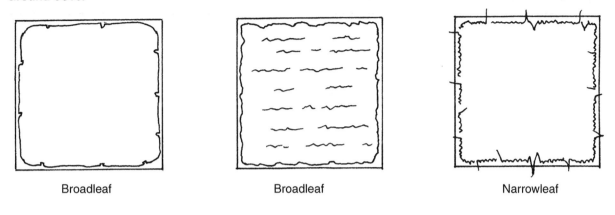

Broadleaf   Broadleaf   Narrowleaf

Vines

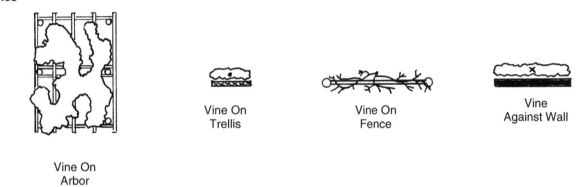

Vine On
Arbor

Vine On
Trellis

Vine On
Fence

Vine
Against Wall

Flowers

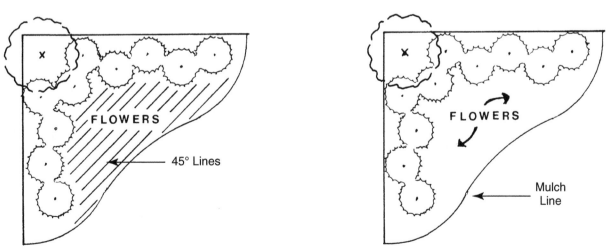

FIGURE 7–1 Symbols for ground cover, vines, and flowers.

NOTES

Exercise 7 Activity

Student Name _____ Date _____ Score _____

Ground Cover

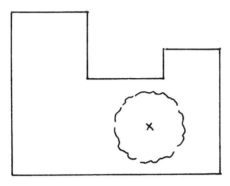

A: Use an appropriate symbol to show broadleaf ground cover in the bed area above.

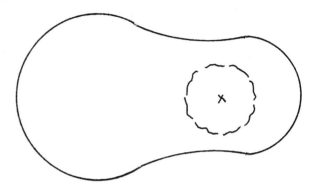

B: Show narrowleaf ground cover in the entire space above.

Vines

A: Draw vines for each of the three sections of fence.

B: Draw a vine to cover the bare wall space above.

Flowers

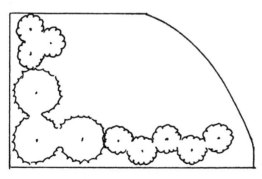

A: Use 45° lines to show a flower space in front of the shrub. Label the space.

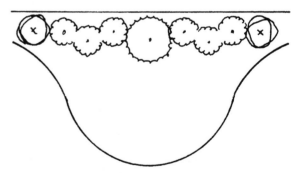

B: Label the bare space as flowers and draw arrows as shown in Figure 7–1.

Evaluation

Consideration	Points	Student Score	Instructor Score
Symbols used are acceptable	30		
45° lines are accurate	30		
Lettering is acceptable	20		
Drawings are neat	20		
Total	100		

Drawing Symbols for Nonplant Features

OBJECTIVE

To provide practice in drawing and understanding nonplant features in the landscape.

SKILLS

After studying this unit, you should be able to:

- Recognize symbols used for nonplant features such as fences, statuary, play equipment, etc.
- Draw symbols used for nonplant features.
- Label nonplant features.

MATERIALS NEEDED

Drawing board and T-square

Drawing pencil (HB, F, H, or 2H)

Eraser and shield

Engineer's scale or architect's scale

French curves

One $8^1/_2 \times 11$–inch sheet of drafting paper

Drafting tape

INTRODUCTION

In previous activities, you learned to draw and symbolize a residence, patios, walks, and plants. There are many other nonplant features and artificial features in landscapes. In drawing symbols for these artificial features, it is important to visualize how the feature(s) would appear from overhead. Imagine you were in a hot air balloon 300 feet overhead. How would the outline of each feature appear?

Artificial and natural items come in many shapes and sizes. Because many have similar shapes, it is important that most of these features be labeled directly on the drawing. The rule of thumb is that if you are in doubt as to whether others will understand what the symbol represents, then you should label it. Smaller size letters are recommended for labeling.

Always draw the features in exact scale. Just as a shrub must be drawn to mature size, nonplant features must be accurate in order to show the exact space that the feature will occupy in the actual landscape. Examine Figures 8–1 and 8–2. Notice the drawing and lettering for each feature. If you don't understand a symbol, ask your instructor for additional explanation.

The samples in Figure 8–1 are drawn on a 1:8 scale, whereas the Figure 8–2 samples are drawn on a 1:10 scale. Use either the 1:8 or 1:10 scale in the activity.

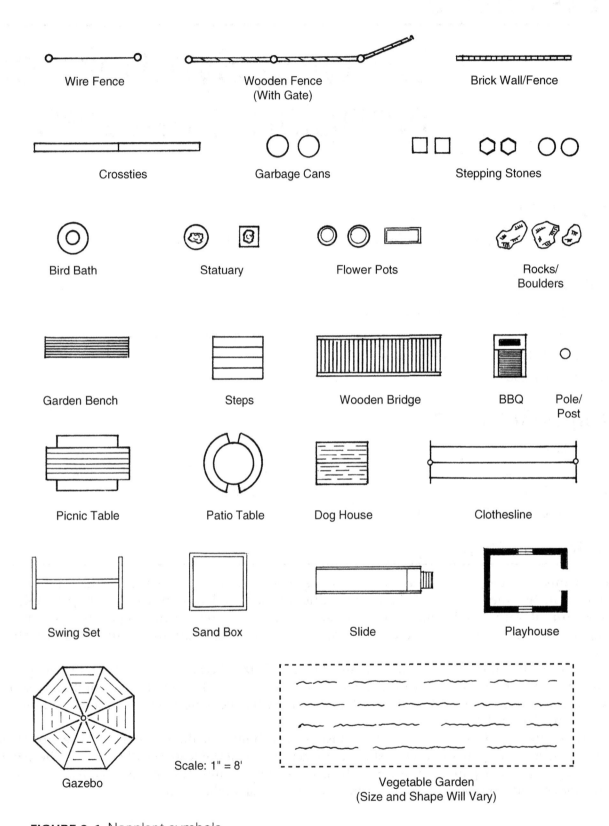

FIGURE 8–1 Nonplant symbols.

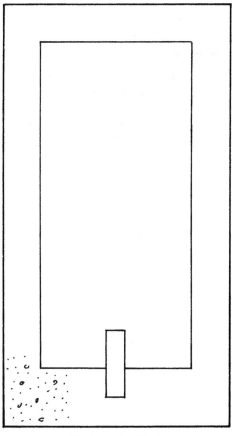

Rectangular Pool
(The Most Common Shape)

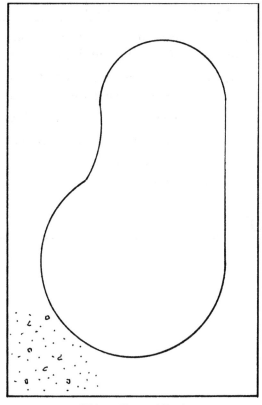

Kidney-Shaped Pool
(Many Custom Shapes
Are Available)

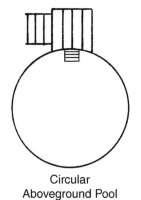

Circular
Aboveground Pool
(Least Expensive However
Least Durable)

Scale: 1" = 10'

Other Recreational
- One-Goal Basketball   40' x 45'
- Tennis   50' x 110'
- Badminton   20' x 45'
- Volleyball   30' x 60'
- Horseshoes   10' x 50'

FIGURE 8–2 Recreational nonplant symbols.

PUT IT INTO PRACTICE

Secure a sheet of vellum, $8^1/_2 \times 11$–inch, and tape it to your drawing board. Use either a 1:8 or 1:10 scale. Please review the Evaluation rubric before completing the activities below.

1. Draw a rectangular flower bed, 16 feet by 24 feet. Enclose the area with crossties (railroad ties) and draw three field rocks (3' to 5' in diameter each) in the flower bed. Place the rocks randomly.

2. Draw a gazebo that is 10 feet wide.

3. Draw a play area 20 feet by 20 feet that contains a swing set, slide, and sandbox.

4. In the remaining space, draw symbols for five or more additional features that one might find in a residential landscape.

NOTES

Exercise 8 Activity

Student Name _____ Date _____ Score _____

Evaluation

Consideration	Points	Student Score	Instructor Score
Drawings are exact scale	40		
Symbols are appropriate	30		
Lettering is neat and accurate	20		
Overall neatness	10		
Total	100		

NOTES

Understanding Foundation Plantings

OBJECTIVE

To understand the importance of and relationships between foundation plantings.

SKILLS

After studying this unit, you should be able to:

- Understand the purpose of and need for foundation plantings.
- Recognize things to avoid in foundation plantings.

MATERIALS NEEDED

Figure 9–1

Pen or pencil

INTRODUCTION

Foundation plantings are those plants used near or against the residence or other buildings. They are used to tie the residence to its outdoor environment, serving as the primary transition in this function. Foundation plants include those plants that are a few feet away from buildings but are a part of the foundation bed area. For example, you might have dwarf shrubs against the building, a bed of ground cover in front of the shrubs, and a small tree growing in the bed area of ground cover. The entire bed area should contain mulch, with a distinctive edge at the point where the bed area and lawn area join.

Foundation plants are the first plants considered in a design. In order to understand foundation plantings, it is necessary to understand the shortcomings of earlier years. Earlier in this century, the public's knowledge of landscaping was limited. Bushes were often chosen for some desirable feature, such as flowers, or some practical use, such as edible fruit, with little regard for mature size, deciduous or evergreen type, or relationship to other plants. As many people have become more affluent and have located in suburban communities, greater emphasis has been placed on designing landscapes that are attractive, functional, and low maintenance.

Figure 9–1 presents some of the more common deficiencies in design. A common shortcoming is the **"toy soldier" effect**. This landscape type uses one species of landscape plants, often round, which are spaced equally with noticeable gaps between plants. This type of design is monotonous, and lacks creativity.

The **overgrown effect** involves the use of plants that are too large for the rooflines or windows of the residence. They dwarf the home and require much maintenance to control their size.

The **crowded effect** involves a large mass of confusion as a result of plants being placed too close at the time of planting. This is done to give the landscape an instant "fullness," and plants eventually lose their individual identity.

The **clipped effect** occurs when all plants are given a regular "haircut" and maintained with a smooth edge. This results in many plants losing their unique growth habits.

The **unbalanced effect** results when too many plants, or larger plants, occur on one side or at the end of the planting. The landscape appears tilted and is obviously out of balance.

Finally, the **hedge effect** occurs when foundation plants are trimmed to a continuous box shape. This lacks in variety, and it gives the foundation no relief from horizontal lines. Hedges are more appropriately used in borders as a living fence.

PUT IT INTO PRACTICE

1. Study the written material and pictures until you understand the concepts.
2. Complete the Exercise 9 Activity without the assistance of the written materials or pictures.

NOTES

"Toy Soldier" Effect

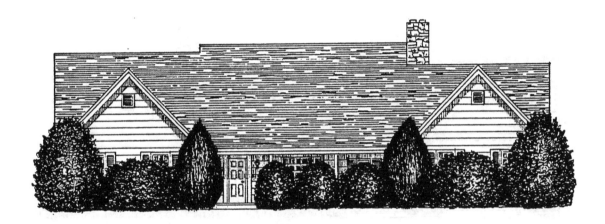

Overgrown Effect

Crowded Effect

FIGURE 9–1 Examples of poorly planned foundation plantings.
(continued on next page)

Clipped Effect

Unbalanced Effect

Hedge Effect

FIGURE 9–1 Examples of poorly planned foundation plantings.
(continued from previous page)

Exercise 9 Activity

Student Name _____ Date _____ Score _____

Fill in the Blanks

Fill in the blanks with the best answers.

1. Foundation plants serve as a transition between the residence and the _____ _____.

2. Foundation plants may extend several feet from the residence if they are a part of the foundation _____ _____.

3. Modern landscapes should be designed so that they are _____, _____ _____, and _____ _____.

4. In the _____ effect, foundation plants are trimmed into a continuous box shape.

5. A landscape appears tilted in the _____ effect.

6. When plantings are _____, the plants lose their individual identity.

7. Selection of plants with no attention to mature size can result in the _____ effect.

8. The hedge effect gives no relief from the _____ lines of the residence.

9. The clipped effect can destroy the unique _____ habit of individual plants.

10. Using one species of plant, spaced evenly, with gaps between, is called the _____ _____ _____ effect.

11. Plants are given a regular, smooth "haircut" in the _____ effect.

12. _____ _____ are the first plants considered in a landscape.

NOTES

Recognizing Faulty Foundation Design

OBJECTIVE

To recognize errors in design and to propose solutions.

SKILLS

After studying this unit, you should be able to:

- Understand solutions to design shortcomings.
- Give one or more names for each area of the design.

MATERIALS NEEDED

All figures from Exercises 9 and 10

Pen or pencil

INTRODUCTION

The material included here will help you understand how to overcome the faults that were studied in Exercise 9, as well as giving you a few new concepts. In designing the foundation planting, one should strive for attractiveness while ensuring that the landscape serves useful purposes and is easy to maintain. Some design principles for foundation planting are listed here:

1. Use taller plants on corners to "soften" the vertical lines while giving relief from the horizontal lines of the residence. The rule of thumb is that the corner plant(s) should mature at no more than half to three-quarters the height of the corner. This is measured from ground level to the lowest point of the "boxing" or overhang. With a one-story house on a flat lot, this will average 8 to 10 feet. Houses with higher foundations will, of course, have taller corners. Two-story homes vary greatly but usually can handle taller plants.

   Normally, medium-size shrubs of 4 to 6 feet in height can be used on corners of one-story residences; large shrubs of 7 to 12 feet can be used on the corners of two-story residences. Following these guidelines will help avoid the overgrown effect.

2. Use lower-growing or dwarf plants under windows. For most windows, a dwarf shrub of some species will provide foliage without resulting in the overgrown effect. In some instances, low windows can be planted with low-growing ground covers, which vary from a maximum of 2 feet down to only a few inches. Another advantage to such treatment of windows is that it ensures that there will be variety in the plants.

3. Show plants on the drawing at mature size (for mature sizes, see Figure 6–1). In drawing the plants, allow them to barely touch or almost touch. This gives a mass effect without being over-crowded. In addition, this avoids the "toy soldier" effect illustrated in Figure 9–1.

4. Maintain balance. If the home or part of the home is symmetrical, both sides of the symmetrical parts should be landscaped alike. (Symmetrical means that all features and measurements are identical on either side of the front door, which is centered.) If the home is unequal or asymmetrical, try to use equal amounts of foliage (leaves) on each end or side of the home. If the home is heavier on one end because of a shorter roof or recessed walls on the other end, use a larger amount of foliage on the light end in order to bring the overall effect into balance (Figure 10–1).

5. If more than one row of plants is used in the foundation planting, place the taller-growing plants nearest the building and place the lowest-growing plants at the front. This will prevent plants from being hidden in the planting.

6. Always use evergreen plants near the wall. Deciduous plants can be harsh when foliage is absent, and planting them near a wall would exaggerate the harsh effect.

7. Use long, sweeping curves for borders between lawn and planting beds. Avoid choppy curves that are difficult to mow and look confusing.

8. Use three or more heights of plants in the foundation planting. This will give variety and a more interesting design while avoiding the hedge effect.

9. Use medium or taller growing dwarf plants on either side of the entry to focus attention on the entry.

10. Repeat some of the same plants on each half of the residence to give an organized look. However, repeating plants does not mean an exact repetition, except in symmetrical landscapes.

Some of these guidelines may require additional thought and research. Use any reference materials you need for additional study, and ask your instructor for additional help as needed.

PUT IT INTO PRACTICE

1. Examine Figures 10–1 and 10–2. Examine the Exercise 10 Activity and try to identify as many faults as possible in Figure 10–2.

2. Complete the Exercise 10 Activity. Assume that the residence is a one-story home with windows 3 feet from ground level.

NOTES

Symmetrical

Partly Symmetrical—Overall Asymmetrical

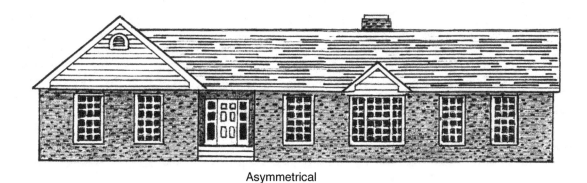

Asymmetrical

FIGURE 10–1 Foundation design requires that balance be maintained.

NOTES

Exercise 10 Activity

Student Name _____ Date _____ Score _____

In Figure 10–2, there are eight or more faults, and you are responsible for finding seven.

To complete the activity, place a number (2 through 8) near the fault, and draw an arrow to the fault. Beside the same number below, describe the fault. In the solution column, describe how the problem can be solved.

One fault has been labeled and explained for you as number 1. Find 7 additional faults to complete the activity, using the example as a guide.

Fault

1. Border between lawn and bed area is too choppy.

2. _____
   _____

3. _____
   _____

4. _____
   _____

5. _____
   _____

6. _____
   _____

7. _____
   _____

8. _____
   _____

Solution

Eliminate small curves. Use long, sweeping curves.

_____
_____

_____
_____

_____
_____

_____
_____

_____
_____

_____
_____

_____
_____

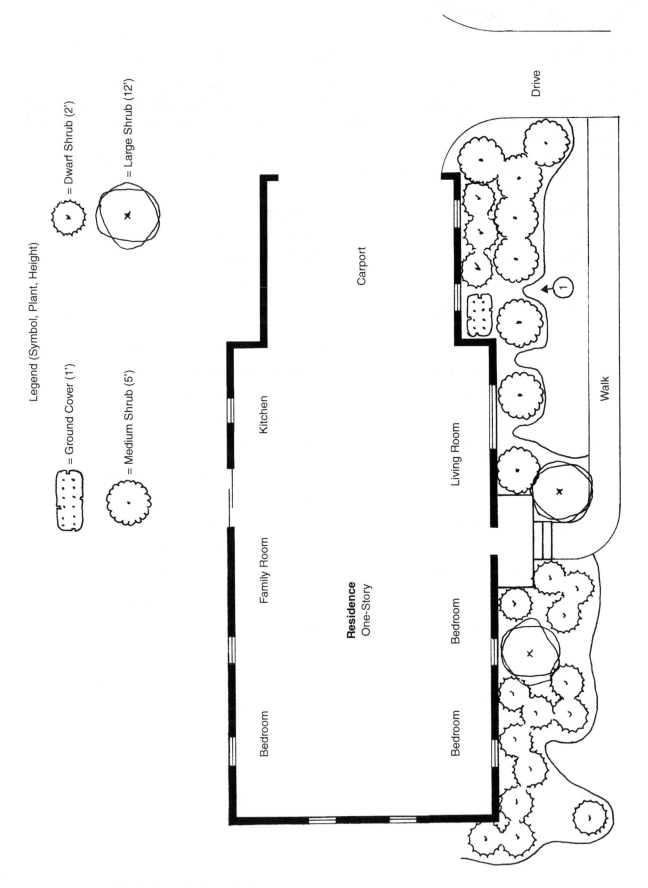

FIGURE 10–2 Faulty foundation design.

EXERCISE 11

Designing a
Foundation Planting

OBJECTIVE

To develop a foundation design.

SKILLS

After studying this unit, you should be able to:

- Apply the principles you learned in Exercises 9 and 10 to develop an acceptable design.
- Understand the relationship of plan view to elevation view.

MATERIALS NEEDED

Drawing board and T-square

Drawing pencil (HB, F, H, or 2H)

Engineer's scale or architect's scale

Eraser and shield

Circle template

French curves

Figures 9–1 and 10–1

INTRODUCTION

In completing this exercise, you will utilize the information presented here, as well as skills from previous exercises, to develop your own foundation design. In designing the foundation, it is important to visualize what the design will look like in real life. For beginning students of design, it is usually helpful to draw an **elevation view** that corresponds to the plan view. The elevation view is simply a front view of the residence as seen from the front yard. This will involve more time, but the time will be well spent as you begin to see the connection between the plan and real life. The sample drawings in Exercise 9 are elevation views. As you gain more experience in design and learn more about individual plants, you will be able to get a mental image of the design without drawing an elevation. Always place plants in the same location on the elevation view as in the plan view, and always make them the same width on each drawing.

At this point, it is unnecessary to give names to the plants you use. It would be helpful, however, to draw a different symbol each time you change plant species. This will enable you at a quick glance to see whether there is variety and also that some species are being repeated. The following guidelines will be helpful as you draft your design. Study the corresponding figures for each step.

Step 1. Always begin your design on the corners. Position the first plant you place so that the central stem of the plant is in line with the corner (Figure 11–1).

A variation of this is to plant three of the same plants on a corner with the middle bisecting the corner angle. If space is not limited, this technique is often used on very large residences where plantings are more numerous to accommodate the larger scale of the structure. In addition, this can help a narrow structure to appear broader (Figure 11–1).

Step 2. Choose the plants that will enhance the entry. They should be taller than the shortest plants used and shorter than the corner plants. When choosing the actual plants, an interesting color, leaf shape, or growth habit will further enhance the entry (Figure 11–2).

Step 3. Fill in the remaining space. Under windows, use dwarf shrubs that will mature at six inches to one foot below the window casing. Often, it is appropriate to use evergreen ground covers as an alternative to shrubs. Larger areas of bare wall space, without windows, can be planted with medium shrubs or larger dwarf shrubs. Sometimes a vine will lend interest to bare wall space (Figure 11–3).

> *NOTE:* One technique often used to enhance the design is to offset or stagger these plants. This also gives more depth to the planting and brings the planting bed out further from the wall for a more landscaped look. In addition, this helps to avoid the formal look that results from straight lines.

Step 4. Add any additional plants desired. This is especially attractive on corners, serving to soften the harshness of corners while giving added variety. Low-growing ground covers can be used as fillers. Shrubs or small trees, carefully placed, can provide additional balance.

The addition of plants, appropriately spaced and of appropriate mature size, greatly improves the appearance and value of the property. However, as previously stated, one must consider the size of the structure and yard. Also, it is best to give some consideration to other gardens in the neighborhood (Figure 11–4).

Side or rear foundations often continue the same technique. This might be important to an end or side that is highly visible. However, the rear landscape is strictly for family enjoyment and will vary greatly from family to family.

Listed here are some examples of ideas for the rear foundation, other than shrubs:

1. Full-length patios, decks, or terraces
2. Herb gardens or cut flower beds
3. Rose gardens or perennial flower plots
4. Play areas for children
5. Other family enjoyment features that do not block the view of the rear garden

PUT IT INTO PRACTICE

Complete the following design directly in Exercise 11 Activities. Refer to activities in Exercises 5 and 6 for sizes of plants as well as Exercises 9 and 10 for foundation planting background knowledge.

1. Make both plan view and elevation view drawings. Use Activity I if you are using a 1:10 scale or Activity II if you are using a 1:8 scale. Review the corresponding Evaluation rubric before beginning the activity.

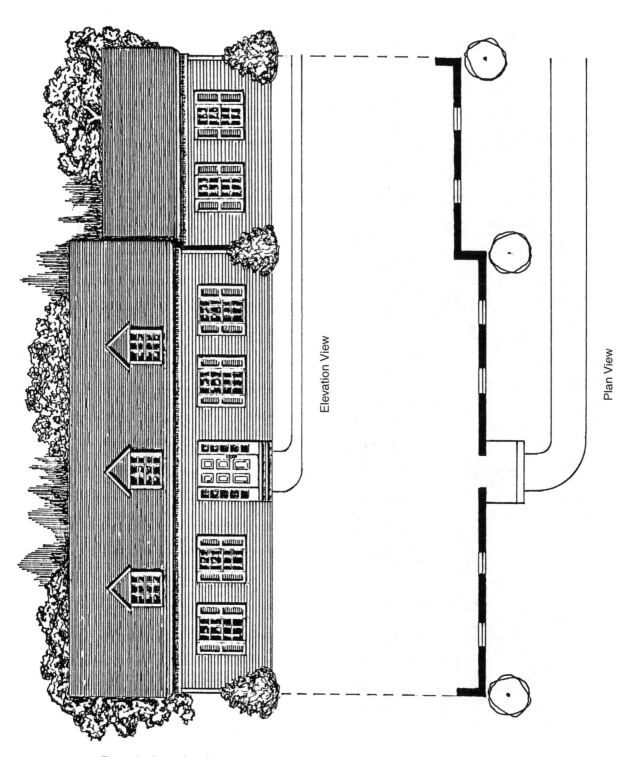

Elevation View

Plan View

FIGURE 11–1 Foundation planting step 1.

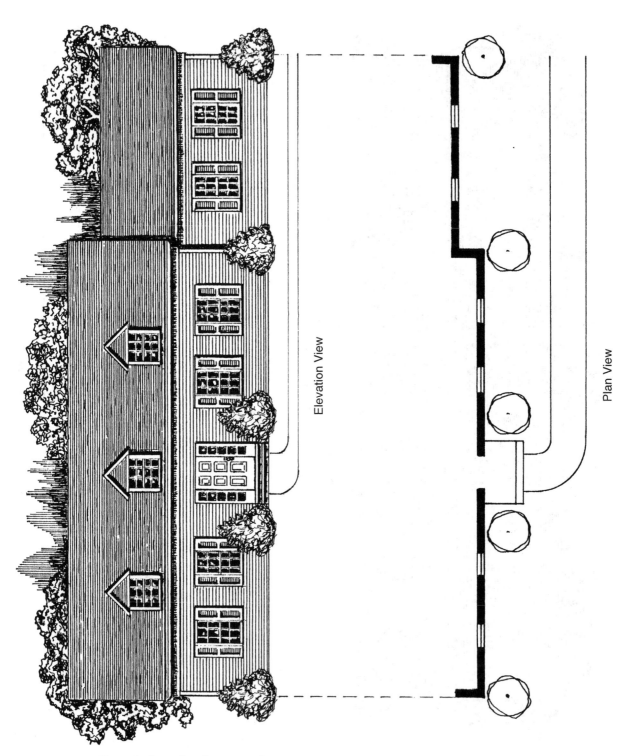

Elevation View

Plan View

FIGURE 11–2 Foundation planting step 2.

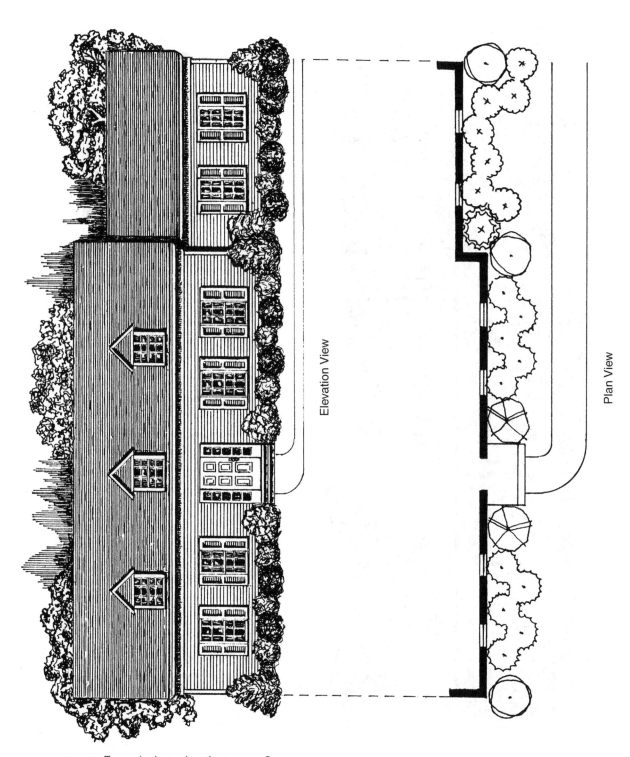

Elevation View

Plan View

FIGURE 11–3 Foundation planting step 3.

Elevation View

Plan View

FIGURE 11–4 Foundation planting step 4.

Exercise 11 Activity I

Student Name _____ Date _____ Score _____

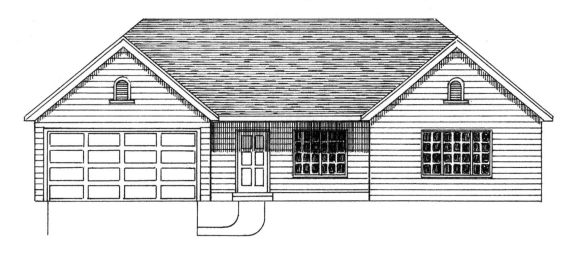

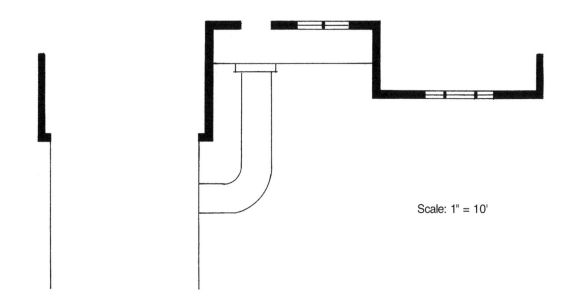

Scale: 1" = 10'

Evaluation

Consideration	Points	Student Score	Instructor Score
Plants are drawn to mature scale	20		
Medium shrubs are on corners	20		
No large gaps exist	15		
All plants are of the correct height	15		
Design is balanced	30		
Total	100		

Exercise 11 Activity II

Student Name _____ Date _____ Score _____

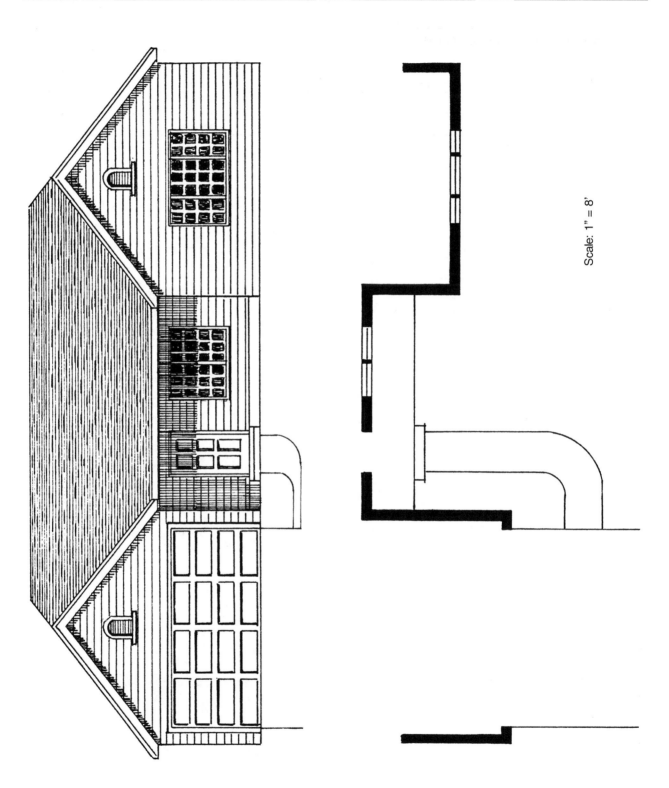

Scale: 1" = 8'

Evaluation

Consideration	Points	Student Score	Instructor Score
Plants are drawn to mature scale	20		
Medium shrubs are on corners	20		
No large gaps exist	15		
All plants are of the correct height	15		
Design is balanced	30		
Total	100		

EXERCISE 12

Organizing Space in the Landscape

OBJECTIVE

To understand the reasons and benefits of well-planned organization of space in the residential landscape.

SKILLS

After studying this unit, you should be able to:

- Understand the need for organizing space.
- Give one or more names for each area of the landscape.
- Identify features for each area.

MATERIALS NEEDED

Figure 12–1

Pen or pencil

INTRODUCTION

One of the more noticeable shortcomings of an average landscape is lack of planned organization of the space. This lack of organization can lead to unnecessary inconvenience, less functional space, and lack of attractiveness. As previously stated, a landscape should be both useful and attractive. Part of the challenge to landscape design is to maximize attractiveness after the usefulness has been planned. In order to understand usefulness and attractiveness, it is helpful to divide the landscape into functional units (areas). Refer to Figure 12–1 to help you define the following areas.

The **public area**, or entrance area, is the area(s) from which passersby will view the residence. This is almost always the front yard. Often, a side yard is part of this area, especially on some corner lots. If a family is limited on funds for landscaping, the public area is usually given the greatest priority. Almost everyone wants to give a favorable public impression.

The public area should be simple; that is, it should not be complicated. For example, the front walk should lead to the front door without taking visitors on a cross-country hike. Likewise, the guest parking should provide ease in maneuvering while providing easy access to the front entry. The public area includes the lawn(s), foundation plants, walks, and drives/parking. The public area should not

79

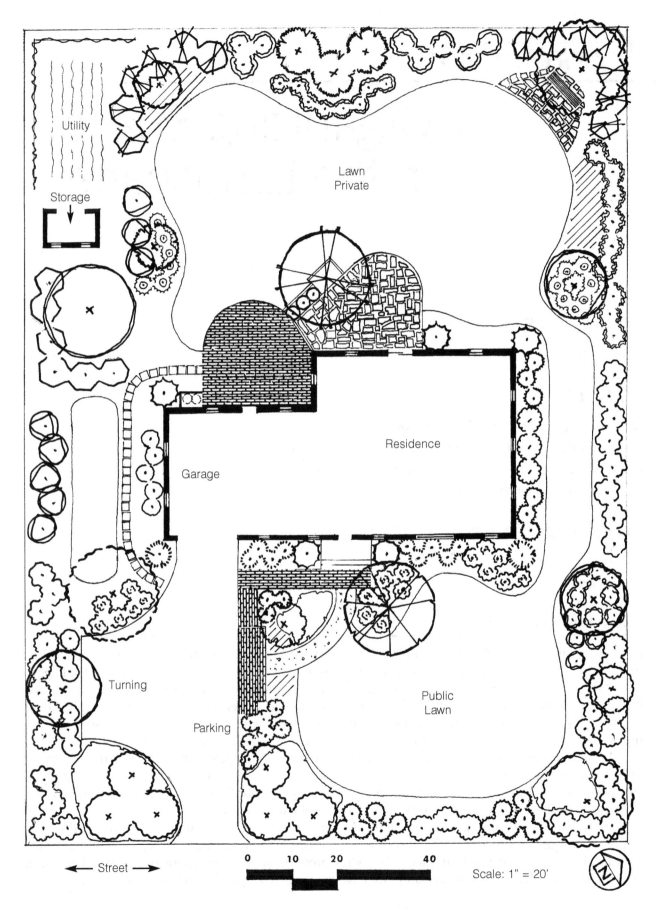

Utility

Storage

Lawn
Private

Residence

Garage

Turning

Parking

Public
Lawn

← Street →

0   10   20   40

Scale: 1" = 20'

FIGURE 12–1 Division of landscape into functional units (areas).

include recreation equipment, play equipment for children, swimming pools, or the like. Avoid the **carnival effect** of objects such as cheaper plastic animals, painted rocks, painted tree trunks, or overuse of weird or unusual specimen plants.

The **private area**, sometimes called the living area, family area, or outdoor living room, is usually located in the backyard or rear garden. Sometimes all or part of a side yard is included. The private area need not be barricaded or a completely isolated area, although it might be. Depending on nature or the interests of the population, dwarf plants may serve simply to identify property boundaries. The purpose of this area is to provide an outside extension of the private living area inside the home—a kind of warm weather retreat.

In addition to people's need for privacy, security is of utmost importance to today's urban population. Plants, fences, or both can be used to make the area private and secure.

The private area should contain the space and arrangement necessary to meet the needs and desires of the family. In its simplest form, the private area will include a deck or patio, an area of open lawn, and plants that provide an attractive view. On a more complex level, the private area can include swimming pools and other athletic facilities, barbecue or picnic facilities, trails, "view" gardens, reflection pools, gazebos, and many other recreation or relaxation features.

The **utility area**, often referred to as the service area or work unit, is also located in the rear garden or side garden. In planning for utility, it is necessary to use the minimum amount of space to fulfill family needs. Such economy of space will free more space for the private area. Some examples of features found in utility areas are vegetable gardens, utility buildings, workshops, pet facilities, greenhouses, clotheslines, compost bins, firewood, fuel tanks, and garbage containers.

The utility area should be screened from the private area, or at least the view should be softened by planting beds. Wherever possible, this area should be located on the driveway side of the yard for ease of access. Time spent designing the layout for the service items will prove highly rewarding as well as visually increasing the size of the remaining spaces.

Finally, some families can benefit from a planned play area for children. Such an area can be planned so as not to detract from the beauty of the private area.

The **play area** should be located adjacent to, or as a part of, the private area. It is critical that the area be visible from the family room or kitchen window, and that there be easy access to a rear entry door. Mulch, fine gravel, sand, or other materials provide more practical surfaces for play areas. Grass is almost always worn and unattractive under play equipment. A play area should contain a variety of playground equipment such as swings, slides, sandboxes, and the like. A shade tree will result in better play area utilization by children during hot weather.

PUT IT INTO PRACTICE

1. Study and review Figure 12–1.
2. Complete the Exercise 12 Activity without the aid of the text or figure.
3. Achieve a score of 80 percent or higher before starting the next exercise.

NOTES

Exercise 12 Activity

Student Name _____ Date _____ Score _____

Fill in the Blanks

Fill in the blanks with the best answers.

1. The three primary areas of a landscape are _____, _____, and _____ areas.

2. A _____ area is needed for families with small children.

3. Dividing the yard into areas encourages the planned organization of _____.

4. A landscape should be both _____ and _____.

5. The front yard is almost always the _____ area.

6. The public area should be _____ rather than complicated.

7. The private area is sometimes called the _____ area.

8. The private area fulfills people's need for both _____ and _____.

9. _____, _____, and _____ are three features found in a private area.

10. The utility area is sometimes called the _____ area or _____ unit.

11. When planning for utility, one should use the _____ amount of space.

12. _____ and _____ are two features that could be found in a utility area of a residence.

13. A swimming pool should be located in the _____ area.

14. The utility area should be located on the _____ side of the yard.

15. A relaxation bench should be located in the _____ area.

16. A small vineyard should be located in the _____ area of the landscape.

NOTES

EXERCISE 13

Making Thumbnail Sketches

OBJECTIVE

To provide practical experience in sketching a property organizational plan.

SKILLS

After studying this unit, you should be able to:

- Plan efficient use of space in a landscape.
- Sketch informal diagrams of landscape areas.

MATERIALS NEEDED

Figure 12–1   Drawing pencil (HB, F, H, or 2H)
Figure 13–1   Eraser

INTRODUCTION

In planning the areas and space in a landscape, it is necessary to plan a specific location for each area and the individual features for each location. Because many changes take place in the development of a final plan, a rough sketch or **property organizational plan** is essential in getting started. This rough sketch does not identify individual plants or shapes for vegetable gardens, swimming pools, and so on. It does identify the area most desirable for each component.

The rough sketch or property organizational plan is commonly called a **thumbnail sketch** because many of the somewhat circular sketches resemble a thumbnail or distorted bubble (Figure 13–1). Another term to use might be **skeleton**, because this drawing does provide the framework from which the completed design develops.

In developing the thumbnail sketch, it is important to use light lines when working with the sheet of vellum that will be the final drawing. In such cases, thumbnails will be erased as the drawing progresses. One method often used is to place another sheet of tracing paper or vellum over your drawing and draw the thumbnails on it. This eliminates any damage to the original. Drawing on the original is far more convenient, provided you are careful.

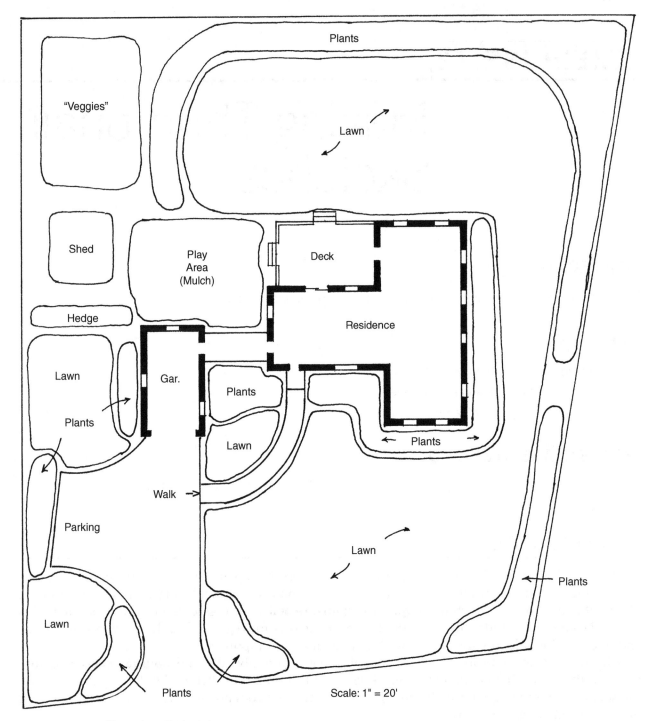

FIGURE 13–1 Thumbnail sketch.

Examine the house and property in the Exercise 13 Activity. Begin visualizing your main area divisions. Before you begin your activity, the following review will be helpful:

1. Draw thumbnail sketches for all surface areas of the property.

2. Include plant beds and lawn areas.

3. Label every thumbnail or bubble.

4. Draw lines that are lighter than normal. If you have trouble sketching with light pressure, a 6H pencil can be used.

5. Remember that exact scale is not necessary for this step. Also, it is not necessary to use guidelines for lettering.

PUT IT INTO PRACTICE

1. Complete a thumbnail sketch for the house and property in the Exercise 13 Activity. Review Figure 13–1 as a guide, and check the Evaluation rubric to familiarize yourself with evaluation criteria.

2. Assume that you will have the following features:

   a. garbage cans

   b. combination workshop/boat shed

   c. swimming pool

   d. vegetable garden

   e. play area for children

   f. firewood stack or pile

   g. neighbors on both sides

3. Notice that the scale is 1:20.

NOTES

NOTES

Exercise 13 Activity

Student Name _____ Date _____ Score _____

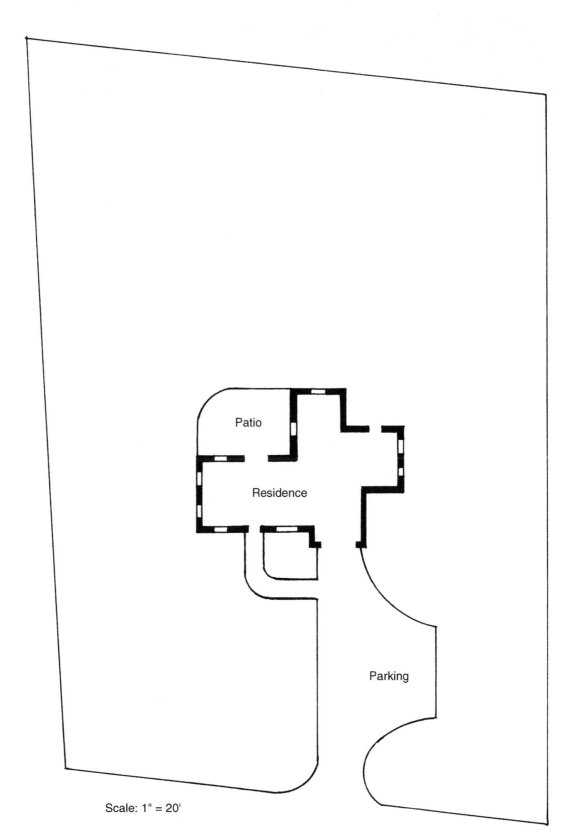

Patio

Residence

Parking

Scale: 1" = 20'

Evaluation

Consideration	Points	Student Score	Instructor Score
All areas are bubbled	30		
Space is divided efficiently	30		
Overall plan is workable	20		
Lines are drawn lightly	20		
Total	100		

Designing Walks and Drives

OBJECTIVE

To provide practice in drawing straight drives and front walks.

SKILLS

After studying this unit, you should be able to:

- Draw a straight drive of appropriate size, length, and material.
- Draw a front entry walk of appropriate size, length, shape, and material.

MATERIALS NEEDED

Drawing board and T-square	Eraser and shield
Drawing pencil (HB, F, H, or 2H)	French curve
Engineer's scale or architect's scale	Figure 4–3

INTRODUCTION

Straight drives are common for houses having carports or garages that open to the front yard. Under almost all conditions, such carports can accommodate a relatively straight drive from the street. Very seldom is a curving drive necessary or desirable. On very large estates, some degree of curve might add interest, if it is not overdone. (Don't take visitors on a Sunday afternoon cruise of your front yard.)

Most garages are either single or double. A rule of thumb is to allow a minimum of 10 feet for each car. For instance, a single-car garage would have a drive no less than 10 feet wide, and a double-car garage would have a drive with a minimum width of 20 feet. Less frequently, you will find garages that accommodate three or more cars. Drives can be made of many different materials. Concrete and asphalt are most common. (Refer to Figure 4–3 for drawing or labeling surfacing materials.) Gravel is a temporary surfacing material and should be treated as such.

There are basically three types of walks that can be used in a residential design. The front walk or **entry walk** is designed to provide a comfortable walking surface for visitors and other guests. In addition, it should serve to direct such visitors to the front door of the residence. It may be straight or it may curve, depending on conditions. For instance, on very small properties where visitors park on the street, a straight walk might be the most practical. However, in most newer subdivisions, the walk runs from the drive to the front door. Some curving is usually necessary. Unnecessary curving will mean unnecessary walking, and it will tempt others to walk on the lawn. The front walk should be a minimum of 4 feet wide to allow two people to walk side by side. A variety of surfacing materials are available. Concrete is most common.

A **secondary walk** is a walk designed primarily for family members. It is placed near side or rear entries, where family traffic is heavy. For example, a commonly used technique is to have such a walk from the drive to a rear entry door. This walk can be smaller; however, a minimum width of 2 feet is desirable. Stepping stones are often used for secondary walks.

A **garden path** is usually more attractive in the overall design when it is not made of solid materials. A path made of mulch, lawn grass, or decorative pebbles would be appropriate. Size will vary depending on need, but 3 feet wide should be adequate.

PUT IT INTO PRACTICE

1. Draw (design) a drive for the residence in the Activity. Use Activity I if using a 1:10 scale or use Activity II if using a 1:8 scale. Review the corresponding Evaluation rubric before beginning.

2. Draw a front entry walk that originates at the drive and runs to the front entry. Leave sufficient space for the foundation plants.

3. Draw symbols on a small section of both the drive and walk to indicate surfacing material.

NOTES

Exercise 14 Activity I

Student Name _____ Date _____ Score _____

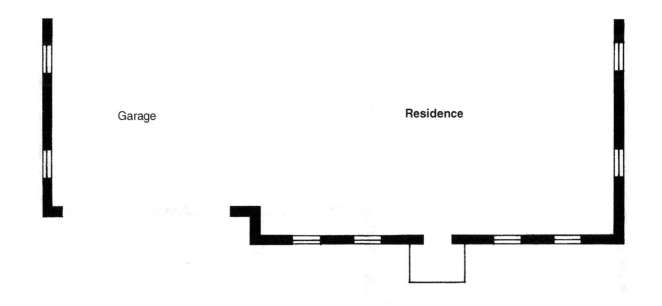

← Street →   Scale: 1" = 10'

Evaluation

Consideration	Points	Student Score	Instructor Score
Drive is of correct size	20		
Walk is of correct width	20		
Surface material is symbolized	20		
Drive and walk are appropriate	30		
Overall neatness	10		
Total	100		

Exercise 14 Activity II

Student Name _____ Date _____ Score _____

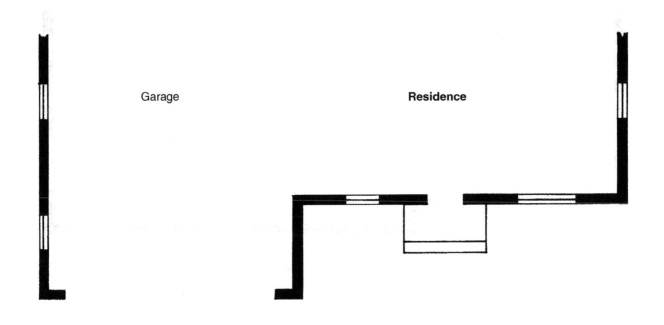

Garage   **Residence**

← Street →   Scale: 1" = 8'

Evaluation

Consideration	Points	Student Score	Instructor Score
Drive is of correct size	20		
Walk is of correct width	20		
Surface material is symbolized	20		
Drive and walk are appropriate	30		
Overall neatness	10		
Total	100		

Drafting Guest Parking and Turnarounds

OBJECTIVE

To provide practical experience in drafting guest parking and/or turnarounds in the landscape.

SKILLS

After studying this unit, you should be able to:

- Understand the need for guest parking and turnarounds.
- Identify different styles of guest parking areas and turnarounds.
- Understand the measurements and angles used for parking and turning.
- Select an appropriate parking and/or turnaround for a given residence, and draft it to scale.

MATERIALS NEEDED

Drawing board and T-square	45-degree triangle or adjustable triangle
Drawing pencil (HB, F, H, or 2H)	French curve(s)
Eraser and shield	Compass
Engineer's scale or architect's scale	

INTRODUCTION

Today, most families very often have two, or even more, passenger vehicles. The need for well-planned space to accommodate family parking has never been greater. Add to this two or three guest cars, and parking becomes even more complicated. Imagine that you need to run an errand and one or two cars have you blocked in. You must either move the cars yourself or have a family member or guest do so. Advance planning could have prevented this situation.

Older neighborhoods, especially those near the inner city, allow for street parking of guests. Such residences have a front walk that leads to the street. More modern suburban developments do not accommodate street parking, and the front walk leads to the drive.

The first consideration is family parking. This might be a garage or a surfaced area near one side that allows quick access to the residence. A multicar family might have both. It is desirable to have a turnaround, where possible, to avoid having to back into the street.

Guest parking should accommodate a minimum of two cars. A socially active family might need more guest parking. Guest parking should always be located in the front yard or one side of the front yard, with a walk that leads guests to the front door.

Examine the examples in Figures 15–1 through 15–4. The measurements given are minimum measurements; it is acceptable to use slightly larger dimensions. Notice that turning areas have an *R* measurement. The *R* stands for radius, and it represents the smallest radius acceptable for turning an average family automobile. Notice the minimum turning radius for a 90-degree turn is 20 feet. To draw a curve of 20-foot *R*, set the steel point and pencil point a distance equivalent to 20 feet on the scale you are using.

Practice drawing a 20-foot *R*. Once you understand Figures 15–1 through 15–4, begin the Exercise 15 activity.

PUT IT INTO PRACTICE

1. For the following activity, if you are using an engineer's scale, use Activity I and a 1:20 scale. If you are using an architect's scale, use Activity II and a 1:16 scale. Review the corresponding Evaluation rubric before beginning.

   a. Draw guest parking and/or turnaround for the residence.

   b. Draw a front walk (4' wide) from the guest parking to the front door.

   c. Complete the drive by connecting it to the street.

   d. Label the areas.

NOTES

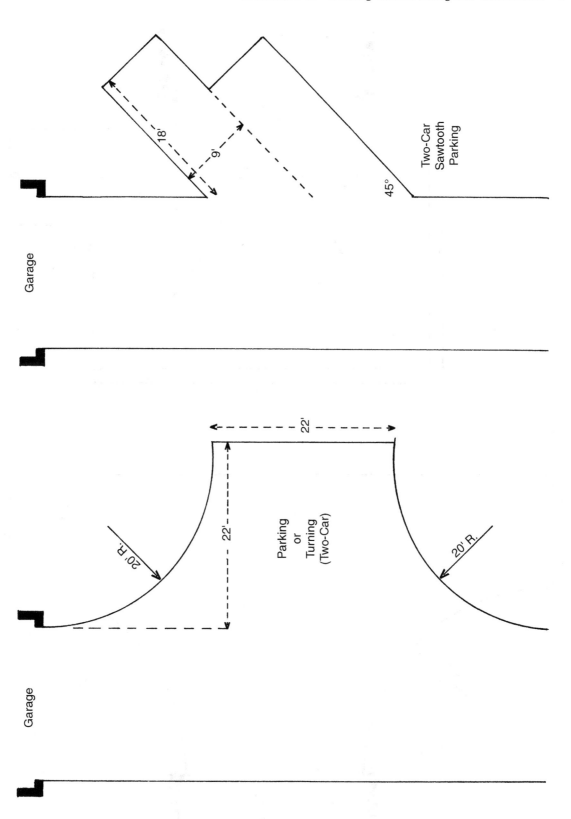

FIGURE 15-1 Garage entry from front.

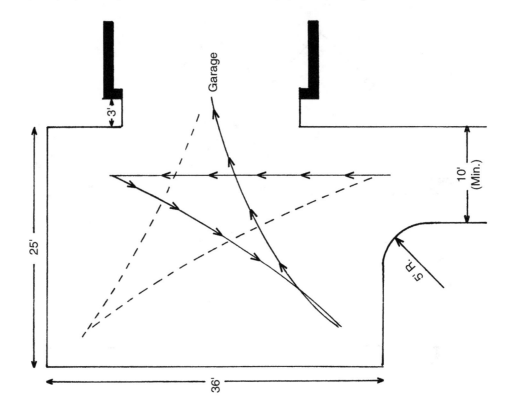

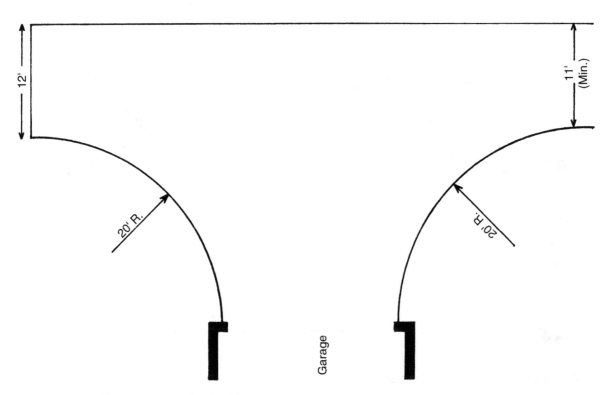

FIGURE 15–2 Garage entry from side.

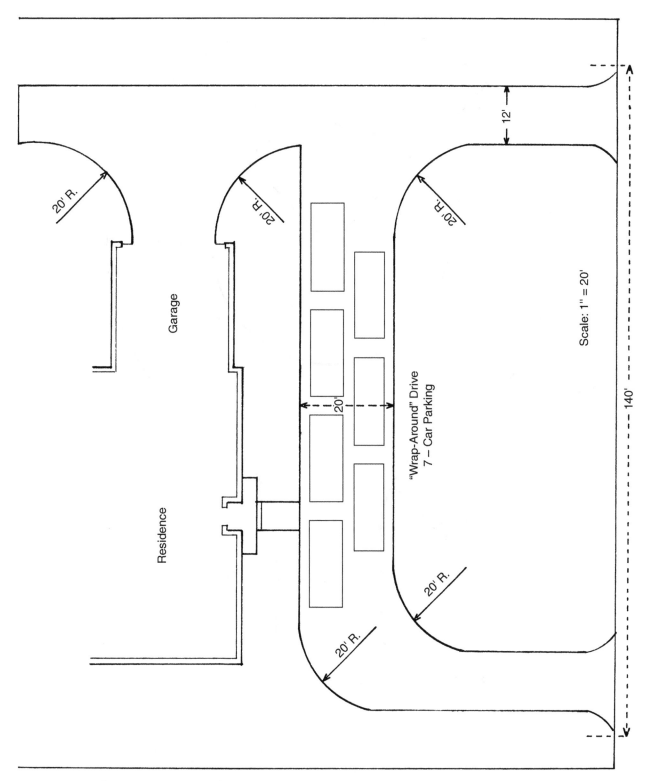

FIGURE 15–3 Large wrap-around.

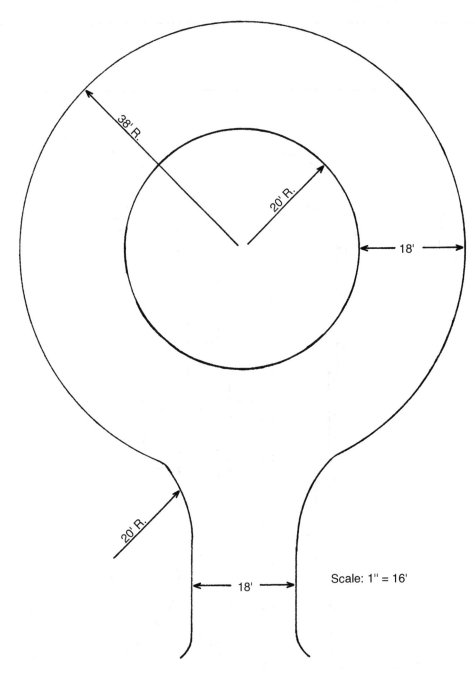

FIGURE 15–4 Small circular drive.

Exercise 15 Activity I

Student Name _____ Date _____ Score _____

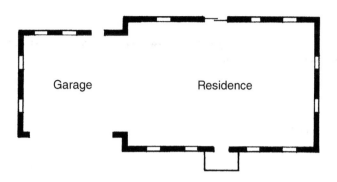

Garage   Residence

←——— Street ———→   Scale: 1"= 20'

Evaluation

Consideration	Points	Student Score	Instructor Score
Measurements are accurate	30		
Parking is appropriate	30		
Angles/radii are accurate	30		
Overall neatness	10		
Total	100		

Exercise 15 Activity II

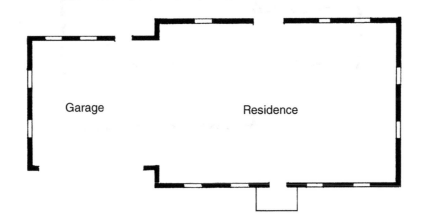

←— Street —→

Scale: 1" = 16'

Evaluation

Consideration	Points	Student Score	Instructor Score
Measurements are accurate	30		
Parking is appropriate	30		
Angles/radii are accurate	30		
Overall neatness	10		
Total	100		

Designing Patios

OBJECTIVE

To provide experience in designing patios for residences.

SKILLS

After studying this unit, you should be able to:

- Understand the need for patios.
- Understand the relationships among size, design, and family needs.
- Design an appropriate patio for a given residence, utilizing appropriate surfacing materials.
- Give reasons for the design and features you choose.

MATERIALS NEEDED

Drawing board and T-square

Drawing pencil (HB, F, H, or 2H)

Eraser and shield

Engineer's scale or architect's scale

Triangle

French curve(s)

Compass

Figure 4–3

INTRODUCTION

A patio is an outdoor surfaced area adjoining (or convenient to) the residence that is used as an outdoor room by family members. The patio is located at the rear of the residence or, occasionally, at one side. Its purpose is to provide an outdoor space for family activities during warm weather. It can be used for many activities, such as barbecuing and dining, socializing, relaxing, sunbathing, and much more. A patio can be partially or totally private, depending upon the desires of the family.

Unlike decks, patios more often are built at ground level, thus providing a kind of transition area between the residence and rear garden. The exact location along the rear of the home is dependent upon rear entrances. One or more entrances to the patio, with a minimum of steps, will greatly enhance usage by family members.

Most modern homes, unless custom built, come complete with a patio or deck (decks are considered in Exercise 17). A patio is usually rectangular and made of concrete, although many different surfacing materials are available. Contractors tend to build patios that are smaller than desirable. Although a patio can be almost any size, it should be no smaller than an average room. Of course, the size can be expanded in future years. It would be helpful to take future expansion into consideration when planning the landscape.

The shapes of patios vary greatly, although, as previously stated, most are rectangular. Almost any geometric shape, or combination of shapes, can be used. In addition, curvilinear or free-form shapes are well suited to some designs. Be careful not to let shape detract from the usefulness or convenience of the patio.

Surfacing should be of permanent, easy-to-maintain material. The most commonly used surface materials are concrete, brick, and flagstone. Concrete is usually the least expensive, whereas flagstone is the most expensive.

The following suggestions should prove helpful in planning the patio:

1. Decide in advance the furniture desired, such as tables, lounge chairs, barbecue grills, and the like.

2. Plan a size that will accommodate family needs.

3. Determine the most convenient location.

4. Choose a surfacing material that is durable and has a nonslip surface when wet.

5. Decide the degree of privacy desired, and plan for any shrubs, fence sections, and so on, that are needed.

6. Determine the degree of bright sunlight during hours of peak usage. Plan for any shade trees or arbors that might be needed.

7. Consider night usage and whether additional lighting or insect control will be needed.

Examine Figure 16–1 and then complete the activity for this exercise.

PUT IT INTO PRACTICE

For the following activity, design a patio for a residence. Use Activity I for a 1:10 scale or Activity II for a 1:8 scale. Show any permanent features such as benches, barbecue pits, and the like. Review the corresponding Evaluation rubric before beginning.

   a. Using the procedures and symbols from Figure 4–3, indicate the surfacing.

   b. Be prepared to defend the rationale (reasoning) for the size, shape, location, and any features that you choose. Present your reasons orally or in written paragraph form.

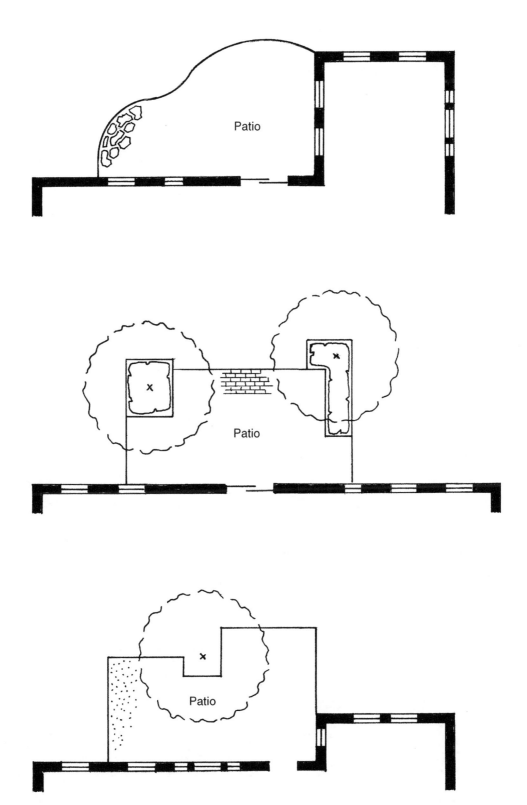

FIGURE 16–1 Examples of patio designs for residences.

NOTES

Exercise 16 Activity I

Student Name _____ Date _____ Score _____

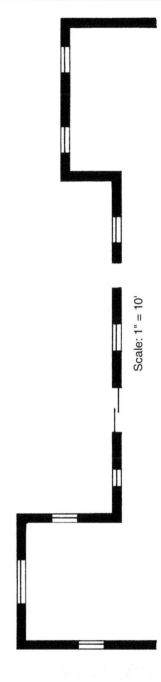

Scale: 1" = 10'

Evaluation

Consideration	Points	Student Score	Instructor Score
Design is feasible	30		
Surfacing is symbolized	20		
Rationale is acceptable	30		
Labeling is accurate	20		
Total	100		

Exercise 16 Activity II

Student Name _____ Date _____ Score _____

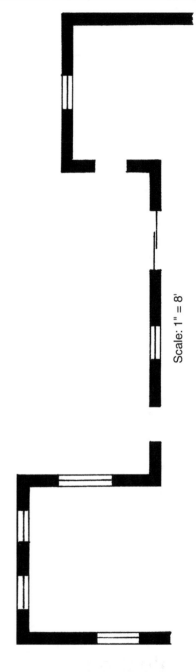

Scale: 1" = 8'

Evaluation

Consideration	Points	Student Score	Instructor Score
Design is feasible	30		
Surfacing is symbolized	20		
Rationale is acceptable	30		
Labeling is accurate	20		
Total	100		

Designing Decks

OBJECTIVE

To provide experience in designing decks for residences.

SKILLS

After studying this unit, you should be able to:

- Understand the need for decks.
- Understand the relationship between design and family needs.
- Design a deck floor plan for a given residence.
- Give reasons for the design and any features you choose.

MATERIALS NEEDED

Drawing board and T-square	Engineer's scale or architect's scale
Drawing pencil (HB, F, H, or 2H)	Triangles
Eraser and shield	Compass

INTRODUCTION

A deck is an outdoor area adjoining the residence that, like a patio, serves as an outdoor space for family activities. However, the deck is made of wood and raised above the immediate ground level. The deck is really a structure and as such requires a structural support system that can range from simple to complex, depending upon the details of the design.

Structural design is not considered in this unit. Those students desiring to learn more about the structural design of decks should consult a local building materials supply house. Employees there can usually give advice on standard structural support specifications. Many such businesses stock brochures and booklets regarding deck support systems. Your project in this unit involves the surface design of the deck.

A well-designed deck can be an addition of great beauty to the residence. However, one must be careful to place usefulness as the primary consideration. One advantage of a deck is that it may exist on more than one level, to accommodate upper levels of the house or sloping terrain.

One disadvantage is that steps must be provided to access the rear garden. The location(s) of access steps should be well planned.

The deck should be built of treated lumber or decay-resistant wood such as redwood. The pattern of the wooden boards can be part of the overall design and add additional beauty. Permanent benches can be added as a part of the design and constructed of the same materials. Railing is essential because the deck is raised above ground level. The verticals of the railing should be close enough to prevent young children from getting through.

Hot tubs or aboveground pools are often added, and they are great complements to a raised deck. Many fine residential designs combine both a deck and patio. This approach is not necessary for this exercise; however, such a design may be a consideration for your final design project.

Examine Figure 17–1 before beginning your design. Note that the boards are shown and constitute part of the overall design.

PUT IT INTO PRACTICE

For the following activity, use Activity I for a 1:10 scale, or use Activity II for a 1:8 scale. Review the corresponding Evaluation rubric before beginning.

a. Design a deck for a residence. Show the boards as 12-inch (1') boards.

b. Show any permanent features such as railings, benches, barbecue pits, spas (hot tubs), and steps.

c. Be prepared to defend the rationale (reasoning) for the size, shape, and location of the deck, as well as any additional features you choose. Present your reasons orally or in writing.

NOTES

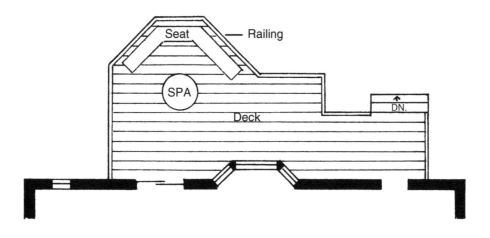

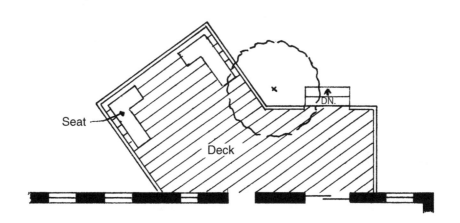

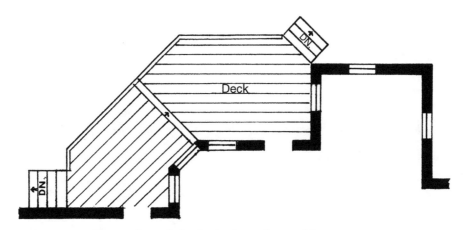

FIGURE 17–1 Examples of deck designs for residences.

NOTES

Exercise 17 Activity I

Student Name _____ Date _____ Score _____

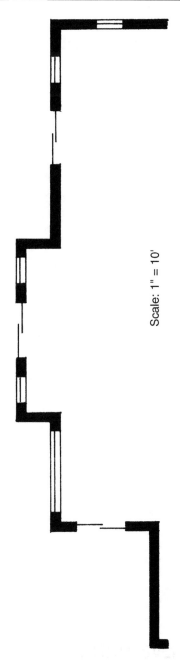

Scale: 1" = 10'

Evaluation

Consideration	Points	Student Score	Instructor Score
Design is feasible	30		
Boards are drawn accurately	20		
Rationale is acceptable	30		
Labeling is accurate	20		
Total	100		

Exercise 17 Activity II

Student Name _____ Date _____ Score _____

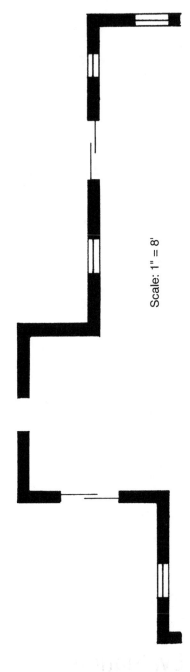

Scale: 1" = 8'

Evaluation

Consideration	Points	Student Score	Instructor Score
Design is feasible	30		
Boards are drawn accurately	20		
Rationale is acceptable	30		
Labeling is accurate	20		
Total	100		

Understanding Balance

OBJECTIVE

To provide practical experience regarding balance in a residential landscape.

SKILLS

After studying this unit, you should be able to:

- Understand the meaning of balance in a landscape.
- Understand how symmetrical and asymmetrical balance are achieved.
- Demonstrate an understanding of balance by completing an activity.

MATERIALS NEEDED

Drawing board and T-square

Drawing pencil (HB, F, H, or 2H)

Eraser and shield

Engineer's scale or architect's scale

Figure 10–1

INTRODUCTION

Balance in a landscape is a condition in which foliage on one side of a view is roughly equal to the foliage on the other side. Balance is often taken for granted when it exists; however, it is noticeable or and even sometimes quite obvious when it doesn't exist. For example, homeowners will often shift or relocate furniture in a particular room until the right balance is achieved. We feel more at ease when things are balanced. Landscape plantings are balanced using foliage mass as a measure of balance. For instance, one small tree might balance three medium-sized shrubs.

The concept of balance may be divided into symmetrical and asymmetrical balance. Symmetrical balance is achieved when the plantings on one side of a view are the exact same as on the other. This type of balance is used when the residence or yard is exactly symmetrical. Symmetrical balance tends to be more formal, so restraint should be exercised.

Asymmetrical balance is achieved when the plantings on each side are not a mirror image of one another, but they do have somewhat equal foliage weight. This type of balance is less formal, and it is the most desirable for the majority of residences.

A more advanced approach to balance, for the experienced designer, considers color, form, and texture. Color is off-balance if, for example, all the plants on one side are red-leaved and all plants on the other side are green-leaved. The same is true for the shape of the plants (form) and the size of the leaves (texture). The Exercise 18 Activity considers just foliage weight in achieving balance.

Examine Figure 18–1 before completing the activity.

PUT IT INTO PRACTICE

1. Complete the symmetrical balance part of the activity. It has been started for you. Fill in a mass of plants along the entire line to achieve symmetrical balance.

2. Complete the activity for asymmetrical balance, using the same procedure as in step 1.

3. Both Figure 18-1 and the Exercise 18 Activity are drawn on a 1:10 scale.

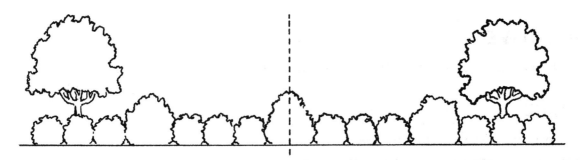

Symmetrical Balance

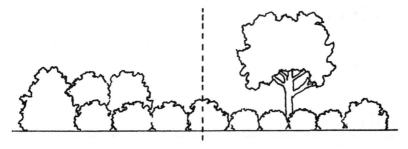

Asymmetrical Balance

FIGURE 18–1 Symmetrical and asymmetrical balance.

Exercise 18 Activity

Student Name _____ Date _____ Score _____

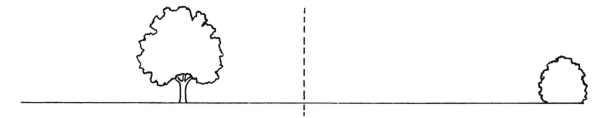

Fill the Blank Space to Achieve <u>Symmetrical Balance</u>.

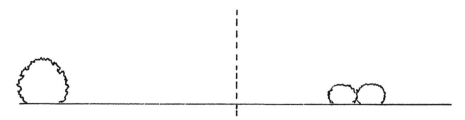

Fill the Blank Space to Achieve <u>Asymmetrical Balance</u>.

Evaluation

Consideration	Points	Student Score	Instructor Score
Balance is achieved	50		
Plantings are to scale	30		
Overall neatness	20		
Total	100		

NOTES

Designing Curvilinear Gardens

OBJECTIVE

To provide practical experience in designing a curvilinear border.

SKILLS

After studying this unit, you should be able to:

- Understand the objective(s) of curvilinear design in a landscape.
- Understand the terms *informal*, *massing*, *variety*, *texture*, and *repetition* as applied to landscape design.
- Complete a design for a backyard border using curvilinear design.

MATERIALS NEEDED

Drawing board and T-square
Drawing pencil (HB, F, H, or 2H)
Eraser and shield
Engineer's scale or architect's scale

French curves
Circle template
Figure 12–1

INTRODUCTION

Just as people vary in their tastes for homes, automobiles, and food, so their tastes vary in plants and their arrangement. For example, at one end of the spectrum is the woodland retreat with no added landscape plantings. On the other end of the spectrum are the formal geometric patterns found in some European estates (e.g., an Elizabethan garden). Somewhere between the two extremes is the design appropriate for most modern residential properties.

The most popular design in residential landscaping is the **curvilinear** design. This concept uses long, sweeping curves to outline planting beds. This approach minimizes abrupt corners and straight lines that are often characteristic of the more formal geometric designs. The curvilinear pattern is considered more informal and invites a more relaxed mood. This is especially important for the family that places great emphasis on a casual lifestyle when at home.

Review Figure 12–1 regarding a curvilinear design for a small rear garden. Notice that the curving lines are long, sweeping lines. These lines are formed where the turfgrass meets the mulch in the planting beds. These lines should be created using tools and maintained effectively for maximum attractiveness. The maintenance of these lines may be the single most important consideration in a beautiful landscape.

Once the general outline is decided, begin designing the plants for the areas. The following guidelines should be considered.

1. **Massing.** This concept has two considerations. First, plants of a kind are grouped together, rather than alternating. Second, plants are placed close enough to look like a mass without overcrowding.

2. **Variety.** Always use several varieties of plants to ensure that the design will be interesting. There should be variations in time of bloom or leaf color and mature height.

3. **Texture.** Texture is most often associated with leaf size. Do not use all coarse-textured or all fine-textured plants. Variation adds interest.

4. **Repetition.** Repeat some plants throughout the design. For example, some of the plants used on one side of the yard should be repeated on the other. In symmetrical designs, the numbers are the same. In asymmetrical designs, it is not necessary to use the same number.

Review Exercise 12 for examples of the previously mentioned concepts, and then complete this exercise.

PUT IT INTO PRACTICE

1. Study Figure 19–1 as an example of a curvilinear design for a small rear garden.

2. *Important:* Turn to the Introduction for Exercise 24 and study the "three magic keys." Use these keys in your design.

3. Complete the Exercise 19 Activity by drafting a curvilinear design for the rear garden illustrated. Be sure to balance the design. Use Activity I for a 1:10 scale, or use Activity II for a 1:8 scale.

NOTES

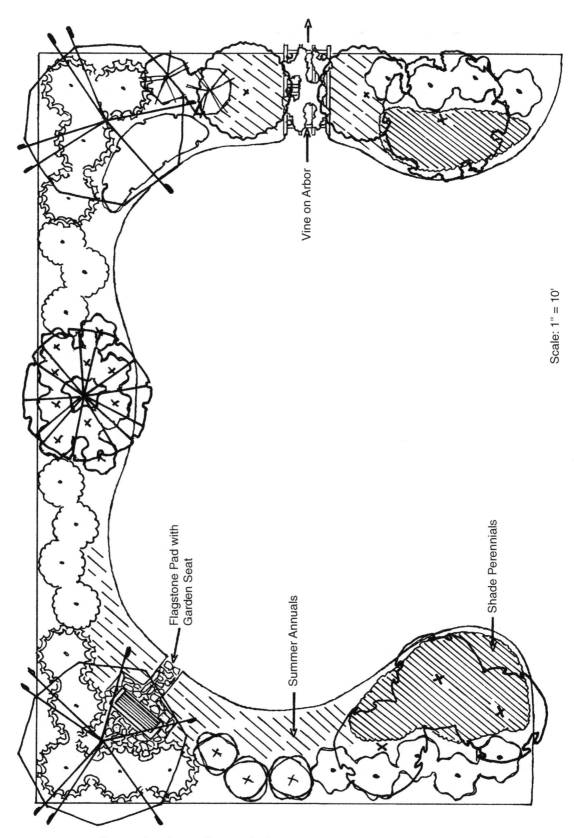

Vine on Arbor

Scale: 1" = 10'

Flagstone Pad with
Garden Seat

Summer Annuals

Shade Perennials

FIGURE 19–1 Example of curvilinear design.

NOTES

Exercise 19 Activity I

Student Name _____ Date _____ Score _____

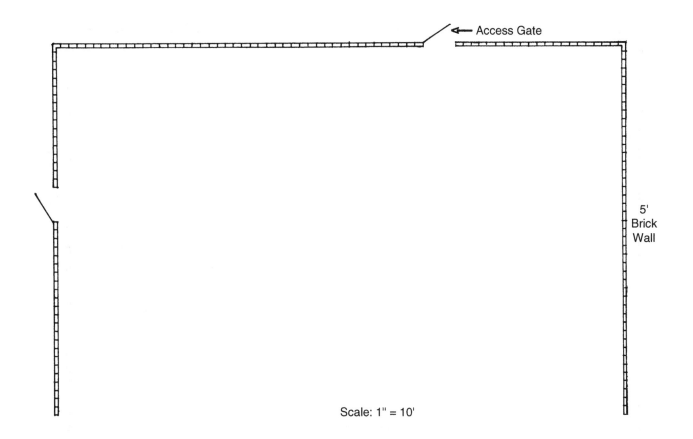

← Access Gate

5'
Brick
Wall

Scale: 1" = 10'

Evaluation

Consideration	Points	Student Score	Instructor Score
Lines are curvilinear	30		
Choppy lines were avoided	20		
Plants are massed	20		
Variety is planned	10		
Repetition is utilized	10		
Drawing is to 1:10 scale	10		
Total	100		

Exercise 19 Activity II

Student Name _____ Date _____ Score _____

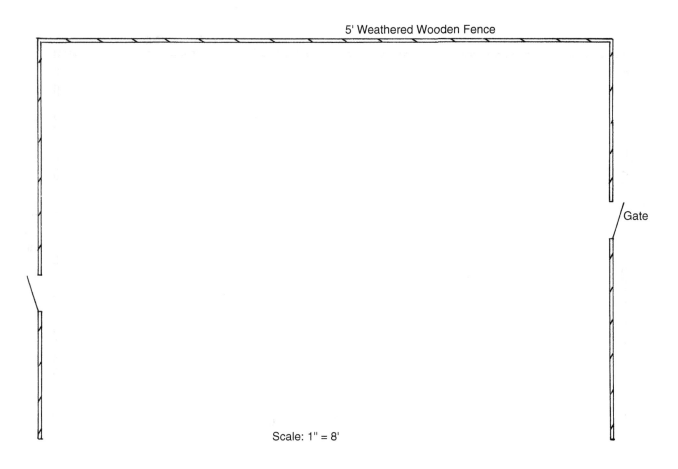

5' Weathered Wooden Fence

Gate

Scale: 1" = 8'

Evaluation

Consideration	Points	Student Score	Instructor Score
Lines are curvilinear	30		
Choppy lines were avoided	20		
Plants are massed	20		
Variety is planned	10		
Repetition is utilized	10		
Drawing is to 1:8 scale	10		
Total	100		

Designing Berms

OBJECTIVE

To help students understand the need for berms in landscapes and provide practice in designing a berm.

SKILLS

After studying this unit, you should be able to:

- State three or more uses of berms.
- Interpret contour lines used in drafting berms.
- Draft a berm to scale and describe its purpose.

MATERIALS NEEDED

Drawing board and T-square

Drawing pencil (HB, F, H, or 2H)

Eraser and shield

Engineer's scale or architect's scale

Drafting vellum or bond copy paper

Drafting tape

Protractor

French curves

Circle template

INTRODUCTION

Berms are artifical mounds of soil that are designed and constructed to serve specific purposes. The size and height of berms can vary greatly. These factors depend on the size of the property, home, and other structures, as well as the effect to be created or the severity of the problem to be dealt with.

The use of well-designed berms in both residential and commercial landscapes has increased dramatically in the past 25 years. However, the percentage of residences having berms remains small. This is most likely the result of a lack of knowledge of berm usage and construction or the costs involved. The beginning designer needs to be familiar with design and construction based on well-defined needs or usage.

Following is a list with the common uses of berms:

1. **To provide privacy.** Many residential properties are too small for privacy berms, so fences, walls, or plants become better choices. However, on larger properties with generous space for family

needs, a combination of berms and evergreen plants can provide privacy and incredible beauty at the same time. Often the need for privacy is so important that homeowners with smaller gardens are eager to sacrifice space for privacy.

One creative use of berms for modern residential communities is the construction of berms on each side of the main entrance. In such cases, berms, in combination with evergreens, convey a feeling of privacy. This is especially true for the first residences on either side of the main entrance.

Some of the largest berms are constructed to soften views and provide privacy for manufacturing facilities, hotels, and apartment complexes. In years past, manufacturing facilities were often eyesores in communities. Unfortunately, many of these buildings are vacant but still exist. However, many modern facilities use well-designed and well-maintained entrance berms that are so attractive the beauty of the community is enhanced.

2. **To reduce noise levels.** Noise reduction is an important use of berms in landscape construction. Some designers claim a reduction of 70 to 80 percent of noise using well-planned berms. This is especially possible when used in combination with evergreens, either broadleaf or narrowleaf (and fences also), to create a layered effect.

   Noise reduction is closely related to providing privacy because many people consider noise an invasion of privacy.

3. **To create a windbreak.** Cold winds can greatly reduce a family's use of the home garden. As with noise, a berm and evergreen plantings can direct wind up and over the home garden while at the same time slowing the speed of the wind. The rear garden (private area), along with frequently used entrances, should receive priority in the design.

4. **To provide interest and variety.** Builders and developers often remove all vegetation and most trees, and they sometimes alter the contour of the land to accommodate more homes of equal value with similar garden sizes. In such cases, one or more small berms, with carefully selected shrubs, ground covers, or flowers, can give the property an interesting character without adding excessive costs or appearing overdone. Such berms can be attractive and need not exceed 18 to 36 inches in height. Small berms are often grassed to suggest a rolling terrain.

5. **To blend the landscape with the environment.** Berms are sometimes used to tie or blend the landscape with the surrounding area. This is especially effective for residences near foothills, mountains, or rocky terrain. A few hardy native plants and a few boulders or large stones can simulate a rocky outcrop or provide a starting point for a natural stream or small waterfall. More attention to detail is usually required to ensure a natural appearance when using this approach.

6. **Others.** There are other conditions for effective use of berms. For instance, a small grass mound near a child's play area is often popular. Children will use the mound in creative ways, including rolling or tumbling downhill on the soft, safe surface. Think of other ways to effectively use a berm in landscape design.

Designing and Constructing Berms

Drafting berms on a landscape plan can be accomplished using the tools learned about in previous exercises.

Landscape designers often design gardens for newly constructed residences. This is ideal, and the designer has more freedom to be creative. However, a large volume of design work involves improvement or add-on work for existing residences. In either case, it is the responsibility of the landscape contractor to supply berms exactly as the designer has planned.

Exact scale is as important for berms as for all other features. Berms should be large enough for the intended purpose without overpowering the landscape.

Once the need for a berm is determined, the designer decides the appropriate shape and the size (length and width). Next, the designer pencils the berm lightly to allow for changes. The finalized shape and size will be the outermost contour line. Berms may be shaped as needed or desired, but the most popular is kidney-shaped or a modified kidney shape. Geometric shapes are seldom used, even in geometric landscapes.

The inner contour lines indicate the height at any given point.

NOTE: Topographic maps used by surveyors and engineers show contours with broken lines for every 2 feet of change in elevation. This landscape designer seldom uses contour lines except when proposing changes. Because berms represent change, solid lines should be used. Also, it is sometimes helpful to show contour lines with spacings of 1-foot change in elevation. The outermost line should be designated as zero (0) because it is the starting point.

Figure 20–1 is a sample berm with contour lines. Elevation is marked on each line. This is the plan view. Elevation views are not required but can be helpful to beginning designers or if requested by a client.

Assume that Figure 20–1 will be a grassed berm to add interest to the property. It is drafted on a 1:4 scale (1" = 4") for visual effect.

Please note the following when studying Figure 20–1.

1. The berm is approximately 25 feet long and almost 15 feet wide at the widest point.

2. Because the berm is grassed, as part of the lawn, the rise is gradual (approximately 23°) to make mowing easier and safer.

3. The berm rises 1 foot for each 2 feet of horizontal distance, and 6 feet of horizontal space is used to attain the maximum height of 3 feet.

4. The berm is asymmetrical since it is taller on one end.

5. The outermost contour line is designated as 0 feet (zero), whereas the inner lines are labeled as 1 foot, 2 feet, and 3 feet.

6. Elevation views, although not always used, allow designers and clients to better understand the effect to be created.

7. The berm in Figure 20–1 is not tall enough, for most gardens, to provide privacy or noise reduction without adding evergreen trees or shrubs.

The berm in Figure 20–2 uses plants to provide greater privacy. After studying this figure, the following observations can be made:

1. This berm is about 43 feet long and 14 feet wide at the widest point.

2. The vertical incline is greater than the berm in Figure 20–1, and it is 1 foot taller to maximize privacy.

3. Contour lines for 1 foot and 3 feet are not necessary because the rise is constant at 45 degrees.

4. The 4-foot berm, combined with large evergreen shrubs, provides over 12 feet of vertical screening.

5. Ground cover was used for ease in maintenance while boulders (optional) were added to make it appear less formal.

6. Heavy mulching will be necessary, especially until the ground cover is well established.

Review Figures 20–1 and 20–2 and the notes for each. Review the uses of berms, and note that there are several other uses not illustrated in this exercise.

Plan View

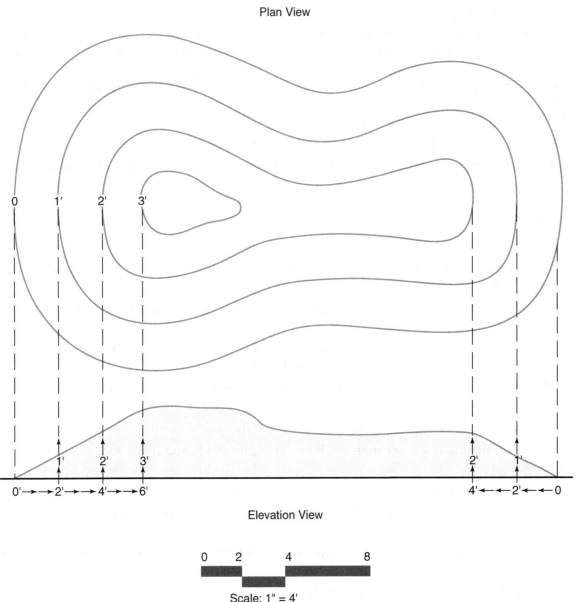

Elevation View

Scale: 1" = 4'

FIGURE 20-1 Sample berm with plan view and elevation view.

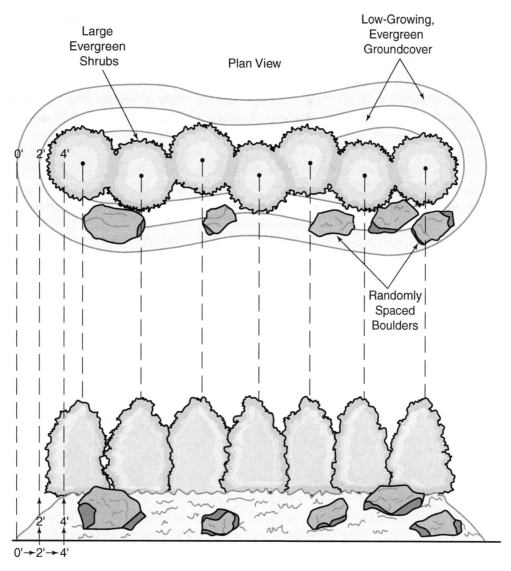

Large
Evergreen
Shrubs

Plan View

Low-Growing,
Evergreen
Groundcover

Randomly
Spaced
Boulders

0' 2' 4'

2'

4'

0' → 2' → 4'

Elevation View

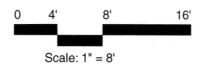

0   4'   8'   16'

Scale: 1" = 8'

FIGURE 20–2 Sample: small privacy berm with plants.

PUT IT INTO PRACTICE

1. Using a piece of drafting vellum, prepare a privacy berm for a hypothetical garden. (No need to draw any features of the residence or property.) Use your scale, 1:10 or 1:8, and draft the berm 40 to 60 feet long and 15 to 20 feet wide at the widest point. The height should be 3 or 4 feet.

2. Briefly describe why the berm is needed. For instance, to provide privacy for a pool, patio, or deck. Perhaps the property borders a busy street. You decide!

3. Draft the outermost contour line and label it 0 feet (zero). Draft and label the other contour lines as illustrated in Figures 20–1 and 20–2.

4. Review Exercises 6 and 7. Using a circle template, draft either medium or large evergreen shrubs as the primary plants. Use ground cover plants or dwarf shrubs in the remaining space.

NOTES

Exercise 20 Activity I

Student Name _____ Date _____ Score _____

Evaluation

Consideration	Points	Student Score	Instructor Score
Reason is given for berm	10		
Accuracy of berm size	15		
Labeled contour lines	20		
Accuracy of elevation view	20		
Accuracy of scale	25		
Drawing is neat	10		
Total	100		

Exercise 20 Activity II

Student Name _____ Date _____ Score _____

Evaluation

Consideration	Points	Student Score	Instructor Score
Reason is given for berm	10		
Accuracy of berm size	15		
Labeled contour lines	20		
Accuracy of elevation view	20		
Accuracy of scale	25		
Drawing is neat	10		
Total	100		

Understanding Focalization

OBJECTIVE

To provide experience in designing focal points for garden areas.

SKILLS

After studying this unit, you should be able to:

- Understand the concept of focalization (focal points).
- Develop and/or identify a focal point in a garden design.

MATERIALS NEEDED

Drawing board and T-square

Drawing pencil (HB, F, H, or 2H)

Eraser and shield

Engineer's scale or architect's scale

Circle template

Triangles

INTRODUCTION

In developing the residential design, it is often necessary to have well-planned areas that immediately catch the eye when entering an area of the garden. Such areas are called focal points, and the concept of planning their use in the landscape is called **focalization**.

Focal points, however important, should be kept to a minimum to avoid competition between the focal points themselves. In the front yard or public area, the primary focal point is the front door. Less dramatic attention can be drawn to the guest parking area or the entrance to the front walk. In the rear garden or private living area, a focal point should provide an attractive view from the patio or deck for the enjoyment of family members and guests (Figure 21–1).

Focalization can utilize specimen plants that are noticeable, natural features such as streams, and/or artifical features such as statuary (Figure 21–2). When plants alone are used, it is possible to shift points throughout the seasons. For example, a flowering tree might serve as the focal point for an attractive area of the garden during spring, then the focus might shift to a grouping of flowering shrubs during summer. Fall color might shift the focus in autumn. Where different focal points occur, it is important that the area attracting attention be neat or worthy of the attention. In addition, it is important that no two plans compete at the same time. One technique of merit is to have plants of varying flowering times in the same bed area. For example, the spring flowering tree might have summer flowering plants growing under it. This will keep the attention on your favorite area.

Of all the techniques involved in the development of the garden plan, the concept of focalization requires as much thought in design as any other concept. Study Figures 21–1 and 21–2, and then complete the Exercise 21 activity.

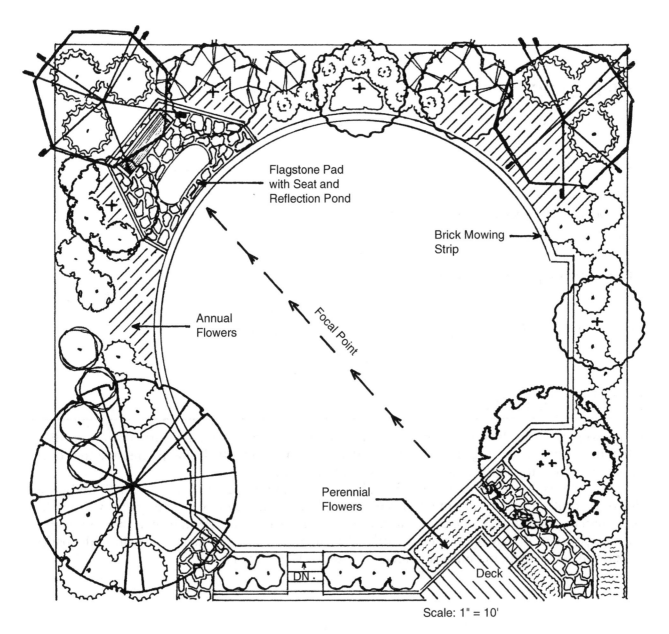

Scale: 1" = 10'

FIGURE 21–1 Focal point in a rear garden or private living area.

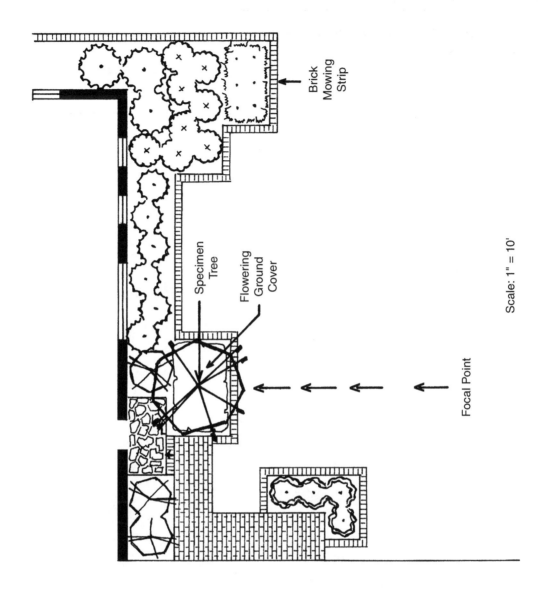

Brick Mowing Strip

Specimen Tree

Flowering Ground Cover

Focal Point

Scale: 1" = 10'

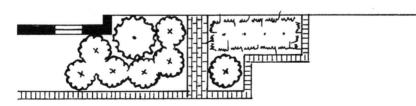

FIGURE 21–2 Focalization utilizing plants and artificial features.

PUT IT INTO PRACTICE

1. Study Figure 21–2, and then draft a focal point for a front entry, using Activity I for a 1:10 scale or Activity II for a 1:8 scale.

NOTES

Exercise 21 Activity I

Student Name _____ Date _____ Score _____

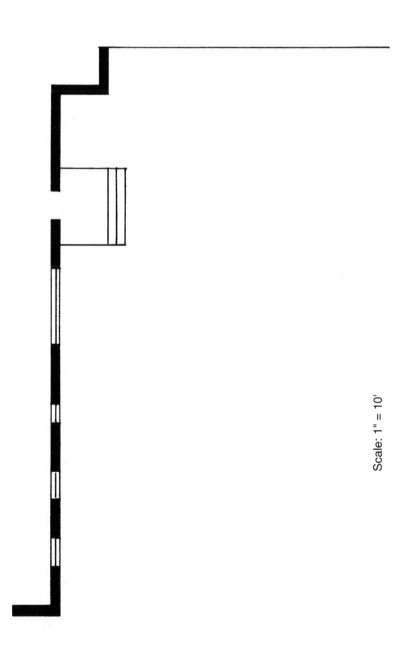

Scale: 1" = 10'

Evaluation

Consideration	Points	Student Score	Instructor Score
Focal points are appropriate	40		
Features are labeled	30		
Exact scale is utilized	20		
Overall neatness	10		
Total	100		

Exercise 21 Activity II

Student Name _____ Date _____ Score _____

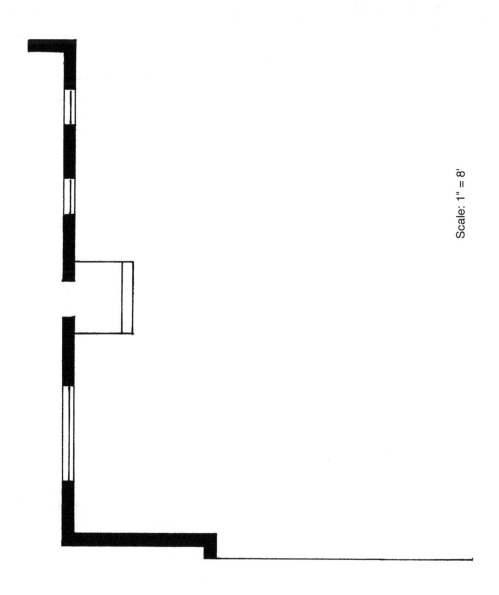

Scale: 1" = 8'

Evaluation

Consideration	Points	Student Score	Instructor Score
Focal points are appropriate	40		
Features are labeled	30		
Exact scale is utilized	20		
Overall neatness	10		
Total	100		

Notice to the Student:

The remaining exercises will constitute your final landscape plan. All remaining exercises will be performed on your sheet of drafting vellum. Use a minimum size of 17" × 22" if using a 1:10 scale or a minimum size of 24" × 36" if using a 1:8 scale.

NOTES

Organizing/ Beginning the Planting Plan

OBJECTIVE

To provide experience in laying out the complete residential landscape planting plan on a sheet of drafting vellum.

SKILLS

After studying this unit, you should be able to:

- Determine the size of paper needed to draft the planting plan.
- Determine the location of and the space needed for the lot, plant list, and title block.
- Prepare the vellum for drafting, and then draft the property for a residence onto the vellum.

MATERIALS NEEDED

Drawing board and T-square

Drawing pencil (HB, F, H, or 2H)

Engineer's scale or architect's scale

Eraser and shield

One sheet of drafting vellum—17 × 22 inches (1:10 scale) or 24 × 36 inches (1:8 scale)

Drafting tape (not masking tape!)

INTRODUCTION

When drafting a landscape plan, it is desirable to have all information for the property, or a designated area of the property, on one sheet of vellum. The information should include the actual design, a plant list, and a title block. The property size and plant list must be considered when deciding the size of the paper needed for the project. It is desirable to use the smallest paper size that will provide adequate space for the project, without overcrowding.

To get started, always have the longest side parallel to the top and bottom edges of the drawing board. Drafting vellum is rectangular in shape and comes in various sizes. The standard sizes are 17 × 22 inches, 18 × 24 inches, 24 × 36 inches, 30 × 42 inches, and 36 × 48 inches. If paper larger than 36 × 48 inches is needed, one should divide the property into areas, such as Area A, Area B, and so on. Alternatively, use a smaller scale, such as 1:20. Scales smaller than 1:16 or 1:20 are not recommended for landscape drafting.

Residential properties are almost always somewhat rectangular in shape; however, many shapes exist. In drafting the property, always draft the front yard adjacent to the bottom edge of the paper and the back yard near the top of the paper. Also, the property should be drafted on the left side of your vellum, with the plant list and title block on the right.

In using the scale, it is easy to determine the size of paper needed. Always allow a minimum of ½-inch space between the border and any edge of the property. A lot 120 feet wide and 180 feet deep would not fit on a 17 × 22–inch sheet of vellum; therefore, a 24 × 36–inch sheet would be needed. A lot 120 feet wide and 150 feet deep would fit on a 17 × 22–inch sheet of paper.

To draft the property, determine the size of paper needed and draw a $^1/_2$-inch border on all four sides. Next, plan for equal "empty" spaces between the front and rear property lines. Last, plan for equal empty spaces between the left border and left property line; between the right property line and plant list; and between the plant list and right border of the paper. Allow around 10 inches of space for the plant list. (Plant lists and title blocks will be considered in future exercises.)

See Figures 22–1A and 22–1B for examples.

Before starting your design, consideration needs to be given to one last feature—a north indicator symbol. This can often be obtained from a plat plan, or surveyor's plan, in the possession of the property owner. If no such plan is available, a simple compass can be used to obtain magnetic north.

The indication of north is important in understanding the patterns of sunlight for the residence and the home garden. If you determine north, you know that west is 90 degrees left, east is 90 degrees right, and south is directly opposite of north (180°). By knowing north, you will be able to predict wind directions prevailing throughout the year that might influence the plants you choose or the location of certain plants. Similarly, you will know where to locate trees that will help to shade patios, decks, and other features at the time of day they are most frequently used by family members. This is often in the afternoon, but it may vary for some families. This should be discussed with the owners during the initial contact.

Another reason for considering direction (north, south, east, west) is the variations in plants' need for light. Some plants are adaptable to either sun or shade, whereas others might need full sunlight or full shade for most of the day. Remember that trees are not the only source of shade. Walls of homes, fences, decks, trellises, and taller shrubs can all affect the amount of light plants receive. As you select plants, be sure to consider light requirements when making selections.

The north indicator can be located in most any prominent place on the drawing. A common place to locate it is on the same line as the scale and scale bar (see Figure 12–1).

Figure 22–2 shows some simple north indicators. You may use one of these, or you can design your own. However, do not overdo it. It should be no larger than the samples shown in Figure 22–2.

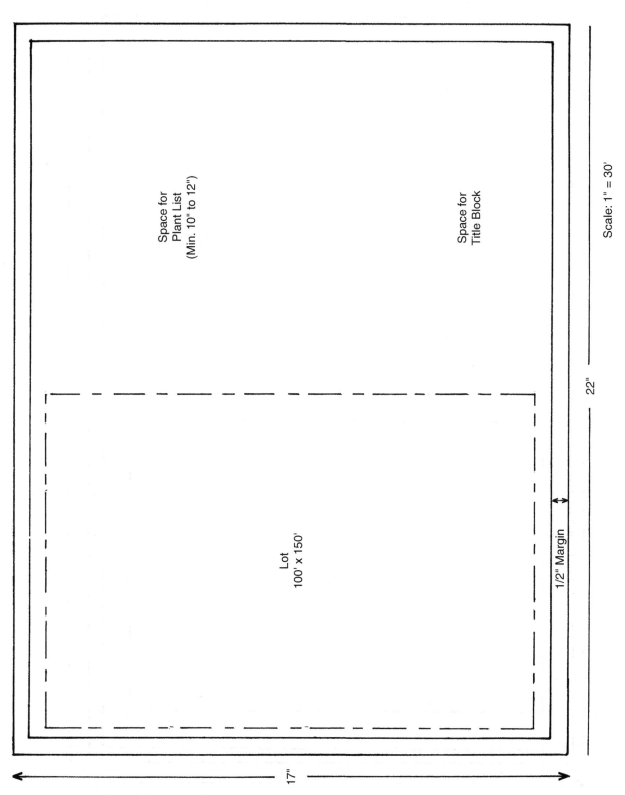

FIGURE 22–1A Example of drafting a property.

NOTE: The scale shown is 1:30 to fit this sample to the page. The plan you create will be much larger.

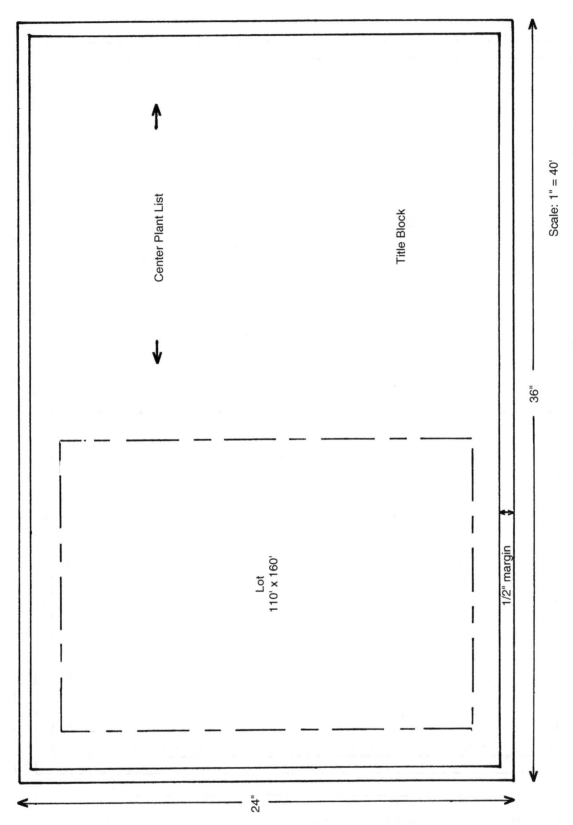

FIGURE 22–1B Example of drafting a property.

NOTE: The scale shown in this drawing is 1:40 to fit this page. Your drawing will be larger.

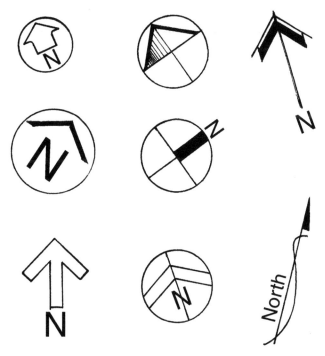

FIGURE 22–2 Examples of simple north indicators.

PUT IT INTO PRACTICE

1. Tape a sheet of drafting vellum to your board. You will draw your complete plan on this sheet. Use a minimum of 17 × 22 inches for the 1:10 scale or a minimum of 24 × 36 inches for the 1:8 scale.

2. Draw a $^1/_2$-inch border around all four sides of the drafting vellum.

3. Draft a property 100 × 150 feet deep for the 1:10 scale or 110 × 160 feet for the 1:8 scale, as shown in Figures 22–1A and 22–1B.

4. Allow a minimum of 10 inches for the plant list, and leave equal spaces between the features and the border.

5. Be sure to include a scale indicator bar and a north indicator symbol.

NOTES

Exercise 22 Activity

Student Name _____ Date _____ Score _____

Evaluation

Consideration	Points	Student Score	Instructor Score
One-half-inch border is used	10		
Sections have equal spacing	20		
Property is to exact scale	40		
10" is allowed for plant list	20		
Overall neatness	10		
Total	100		

NOTES

Placing Artificial Features on the Plan

OBJECTIVE

To provide experience in properly locating buildings and other artificial features on a landscape plan.

SKILLS

After studying this unit, you should be able to:

- Draft a house onto a property according to given specifications.
- Draft a drive, entrance walk, and parking for the residence.
- Draft a patio or deck, access walks, fences, and other artificial features, as desired, onto the plan.

MATERIALS NEEDED

Drawing board and T-square

Drawing pencil (HB, F, H, or 2H)

Eraser and shield

Engineer's scale or architect's scale

Triangles

Compass

French curves

Drafting vellum from Exercise 22

Drafting tape

INTRODUCTION

Wherever possible, it is desirable to have a rear garden equal to or greater than the front yard in size. The reasoning is that the rear garden is where family members will engage in family activities. For most properties, this can be accomplished by locating the front of the residence between 50 and 100 feet from the street. However, the rear garden is sometimes smaller in order to locate the residence an appropriate distance from the street. As a general rule, the distance from the front of the house to the street is reasonably consistent for each house on the street, even though rear properties might vary.

The distance from each side of the house to the property line is important. Always locate the house at least 10 feet from the property lines. When the garage opens to the front, the house can be centered between the side property lines. However, when turnarounds are used to accommodate garage entry from the side, the house may shift toward the property line on the opposite site. Once the location for the house has been determined, it should be drafted onto the property.

As a reminder, in most cases the landscape designer begins to work with the homeowner after the house has been built, and the drive, front walk, and patio/deck have already been constructed. In such cases, guest parking has to be planned utilizing the existing drive. Sometimes homeowners will modify the walk or patio with your help in design.

In this exercise you will landscape a given home, but you will have the flexibility to design the drive, guest parking, walks, patio, and other features. You will need to refer to information from several previous exercises in your design.

PUT IT INTO PRACTICE

1. Draft the house from this activity onto the lot you drew in Exercise 22. Locate the front of the house 50 feet from the property line.

   a. Use your scale, 1:10 or 1:8, and draft the berm 40 to 60 feet long and 15 to 20 feet wide at the widest point. The height should be 3 or 4 feet.

   b. Assume that the house is a one-story residence with all windows 3 feet off the ground.

2. Design a drive, guest parking, walks, and rear patio/deck. Plan for any other desired permanent features, such as a swimming pool, fences, storage buildings, or the like. You should make a very lightly drawn bubble diagram or thumbnail sketch to help you organize areas of the landscape.

   a. Use your scale, 1:10 or 1:8, and draft the berm 40 to 60 feet long and 15 to 20 feet wide at the widest point. The height should be 3 or 4 feet.

NOTES

Exercise 23 Activity

Student Name _____ Date _____ Score _____

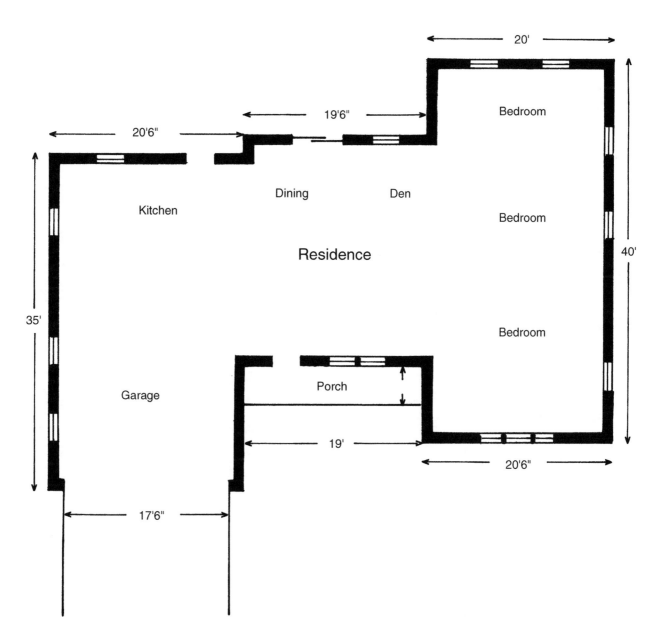

Evaluation

Consideration	Points	Student Score	Instructor Score
House is located appropriately	30		
Walks, drive, etc., are acceptable	30		
Exact scale is maintained	20		
Overall neatness	20		
Total	100		

NOTES

Developing the Design

OBJECTIVE

To develop a total design for a residential landscape.

SKILLS

After studying this unit, you should be able to:

- Utilize previously acquired skills in the development and drafting of a landscape for a residential property.
- Organize the yard into functional areas.
- Label items as per previous instruction.
- Implement the "three magic keys."
- Utilize a checklist in preevaluation.

MATERIALS NEEDED

Drawing board and T-square

Drawing pencil (HB, F, H, or 2H)

Engineer's scale or architect's scale

Eraser and shield

Triangles

Compass

French curves

Circle template

Sheet of vellum used in Exercise 23

Drafting tape

INTRODUCTION

You are about to begin the most time-consuming exercise—the complete landscape design. Care should be used in placing plants because this will determine—more than any one factor—the quality of the design. You may need to experiment, so do not hesitate to use your pencil and eraser in making desired improvements. You should begin with the foundation planting for the residence.

In designing the total yard plan, there are "three magic keys" to an attractive design:

1. Landscape the borders of the property. This is especially important in the rear garden.

2. Leave open areas of lawn. These areas will vary greatly in size from one garden to another, but they are present, to some degree, in the very best landscapes.

3. Use curving lines in the borders (except for geometric designs). The curving design (curvilinear) is far more popular than geometric designs, because it gives a more natural look.

Refer to Figure 12–1 for examples of these keys.

A landscape checklist is provided for your use in evaluating the design phase of your plan. Following is a brief explanation of checklist items. Some items will be a review of previously considered materials.

I. Foundation Planting

    A. Use taller plants on corners. Medium-size shrubs should be used for one-story homes; large shrubs for two-stories or taller.

    B. Use dwarf shrubs or ground covers under windows 4 feet or less above ground level.

    C. Enhance the entry by using noticeable or different plants.

    D. Balance the planting with equal foliage mass. Remember: Symmetrical structures must have the same number and species of plants on each side of the front door.

    E. Repeat some of the same plants on each end, even if it is an asymmetrical landscape.

    F. Variety should be used in selecting plants. This includes texture (leaf size).

    G. Massing of plants will give a more pleasing effect. Plants should be drawn to mature size and allowed to touch, or almost touch, adjacent plants.

II. Overall Design of Property

    A. Balance—The entire foliage mass should appear in balance.

    B. Repetition—Same as for foundation plants.

    C. Variety and texture—Same as for foundation plants.

    D. Massing—Same as for foundation plants.

    E. Simplicity—Absence of inconveniences, such as unnecessary curves in walks, fences without gates, and the like. (Repeating some plants can be included here.)

    F. Circulation—Ease in getting around in the garden. Can you get a truck in the rear garden, if necessary?

    G. Areas defined—Yard is divided into functional areas: public, private, utility, and, if needed, a play area for children.

    H. Focal points—Areas that draw attention, especially in the rear garden and front entry of the residence.

    I. Avoid straight lines of plants, if possible.

    J. Provide guest parking.

    K. Borders of property should be landscaped.

    L. Open areas of lawn should be planned.

    M. Borders should be curving (except in geometric designs).

III. Drafting Skills

    A. Use exact scale.

    B. Letter items that need labeling. Be neat.

    C. Strive for overall neatness.

    D. Lines should be dark and smooth.

    E. Different plant symbols are used for different plant species.

Landscape Checklist

Use this form for preevaluation of your design. Place an "X" in the block beside an item that is acceptable, and correct any deficiencies before completing the exercise evaluation.

Foundation Planting

- ❏ Appropriately sized plants were used on the corners of the house.
- ❏ Dwarf plants or ground covers were used under the windows.
- ❏ Entry to the residence was enhanced.
- ❏ Foliage mass appears balanced.
- ❏ Plants were repeated in the design of the foundation planting.
- ❏ Variety in plants and texture was used.
- ❏ Plants were massed.

Overall Design of Property

- ❏ Overall design of the property appears balanced.
- ❏ Plants were repeated throughout the design.
- ❏ Variety in plants and textures was used.
- ❏ Simplicity was evident.
- ❏ Circulation provides ease in movement throughout the garden.
- ❏ Areas were well defined and appear organized.
- ❏ Focal points exist in both the front and rear gardens.
- ❏ Straight lines of plants were avoided wherever possible.
- ❏ Guest parking has been provided and is appropriate.
- ❏ Plantings have been designed for the borders of the yard.
- ❏ Open areas of lawn have been planned.
- ❏ Borders are curving (or geometrically designed).

Drafting Skills

- ❏ Exact scale was used throughout the design.
- ❏ Lettering is both complete and neatly drafted.
- ❏ Lines are dark, smooth, and consistent.
- ❏ Overall neatness was maintained throughout the entire design.

PUT IT INTO PRACTICE

1. Using the house and property drafted in Exercise 23, lightly sketch bubble diagrams of areas and features. Lines can be erased as you develop permanent lines.

2. Draft the entire plan, using reference material from this exercise and previous exercises.

3. Maintain exact scale for every feature.

4. Keep a separate list of plants used.

NOTES

Exercise 24 Activity

Student Name _____ Date _____ Score _____

Evaluation

Consideration	Points	Student Score	Instructor Score
Foundation planting is acceptable	30		
Overall design is acceptable	30		
Lettering is adequate and neat	10		
Exact scale is used	20		
Overall neatness	10		
Total	100		

NOTES

Selecting Plants

OBJECTIVE

To provide students with experience in plant selection skills

SKILLS

After studying this unit, you should be able to:

- Understand size variations or size groups in mature plants.
- Understand uses of various groups.
- Locate sources of information on plants.
- Prepare an informal list of plants.

MATERIALS NEEDED

Notebook paper and pencil or pen

Drawing from Exercise 24 (for reference purposes)

Various books, magazines, booklets/brochures, and nursery catalogs containing specific information on landscape plants. Usually your local extension service can provide information on plants appropriate for your area. Ask your instructor to provide assistance in locating pictures or specific information on plants.

INTRODUCTION

In the selection of landscape plants, it is necessary to group plants according to the sizes they attain at maturity. For example, dwarf shrubs or large trees are such groups, and each contains many fine plants suitable for landscape purposes. One of the first questions people ask is "How big does it get?" Information on specific plants should include the following:

1. Region(s) in which it grows
2. Size—height and width
3. Information on leaves, flowers, and fruit
4. Light requirements, such as sun or shade

5. Habit of growth or shape of the plant

6. Evergreen or deciduous (loses leaves in fall or not)

The following describes size groups according to height, as well as some possible uses for plants in each group.

Low-Growing Ground Covers—0 to 2 feet

Ground covers are used as a substitute for turfgrasses, where mowing might be difficult; under low-branched trees; in shady areas where grass might not grow; and to add variety to the landscape.

Vines—Size varies

Vines can be trained on walls, arbors, trellises, and fences. Vines are good substitutes for shrubs where space is limited, and they add interest.

Dwarf Shrubs—Less than 4 feet

Dwarf shrubs are used in foundation plantings, especially under windows. They are often used as low-growing hedges or as a substitute for turfgrasses or ground covers. They are best used in masses in bed areas in the landscape.

Medium Shrubs—4 to 6 feet

Medium shrubs are used on corners of one-story structures, unclipped hedges, border plants, and as background for smaller plants.

Large Shrubs—6 to 12+ feet

Large shrubs are used as corner plants for taller structures, windbreaks and hedges, and as tree-formed specimens.

Small Trees—Less than 20 feet

Small trees are used for shade or accent and to give variety. They look nice in groupings and do well in smaller areas, where a large tree would not be appropriate.

Medium Trees—20 to 40 feet

These trees may serve as the large trees for smaller gardens. Some of the best flowering trees are found in this category.

Large Trees—Over 40 feet

Large trees are used as screens, specimen plants (in many cases), and for shade.

Specialty Plants—Great variation

These include ornamental grasses, bamboo, palms, fruit trees, banana trees, and others. These plants have a variety of functions in landscapes. You should study these plants carefully before using them in designs.

PUT IT INTO PRACTICE

1. Prepare an informal list of plant names for the plants in your plan (see Figure 25–1 for an example). You will need both common names and botanical names. This informal list will be drafted onto your plan in the next exercise.

2. A good landscape plan should contain a minimum of 20 different species of plants. Be sure to select 20 or more.

A 2 Red Maple - *Acer rubrum* 'October Glory'

B 1 River Birch - *Betula nigra*

C 2 Eastern Redbud - *Cercis canadensis*

D 1 Yoshino Cherry - *Prunus yedoensis*

E 3 Flowering Dogwood - *Cornus florida* 'Cloud Nine'

F 9 Hybrid Rhododendron - *Rhododendron* x 'Roseum Elegans'

G 7 Manhattan Euonymus - *Euonymus kiautschovicus* 'Manhattan'

H 2 English Holly - *Ilex aquifolium*

I 3 Pampas Grass - *Cortaderia selloana*

J 9 Mountain Laurel - *Kalmia latifolia*

K 14 Kurume Azalea - *Azalea obtusum* 'Hino-Crimson'

L 2 Hybrid Clematis - *Clematis* x *jackmanii*

M 9 Helleri Holly - *Ilex crenata* 'Helleri'

N 193 Japanese Spurge - *Pachysandra terminalis*

FIGURE 25–1 Informal list of plants.

NOTES

Exercise 25 Activity

Student Name _____ Date _____ Score _____

Evaluation

Consideration	Points	Student Score	Instructor Score
List contains common and botanical names	40		
Plants are suitable for your region	40		
List contains 20 or more species	20		
Total	100		

NOTES

Labeling Plants on a Plan

OBJECTIVE

To provide experience in labeling the plants on a landscape plan.

SKILLS

After studying this unit, you should be able to:

- Understand the need for labeling plants.
- Key or code different plants on a landscape plan.

MATERIALS NEEDED

Drawing board and T-square

Drawing pencil (HB, F, H, or 2H)

Eraser and shield

Engineer's scale or architect's scale

Triangles

Drawings from Exercises 24, 25, and 26

Drafting tape

Informal plant list from Exercise 25

INTRODUCTION

In order for a landscape plan to be usable, one must correctly label all plants in such a way that others can read the plan. One system is to write names on the plan near the plant, and draft a line from the name to the plant. On complete plans, this system can become confusing, with a mass of names cluttering the plan.

A proven method is to use keys or codes to identify the plants on the plan. These codes can then be matched to codes in the actual plant list to find the name of the plant (Figure 26–1). Such codes use letters of the alphabet to key the plants and numbers to give the quantity of plants in the group or area.

One approach is to start coding in the front yard and work your way to the rear. Some instructors prefer to list trees first, then shrubs. Give the first plant(s) you identify the code "A." If you have only one A, then label it A-1. If there is a group of plants—for example, five—of the same species, then write A-5. Draw a line to the plant or to the group. The next plant you code will be identified as "B." If your list contains more than 26 different plants, then, after "Z" has been used, starting labeling "AA," "BB," and so on. The codes should be parallel to the bottom edge of the sheet and written in 1/10-inch or 1/8-inch capital letters.

Study Figure 26–1. Have your instructor provide additional explanations, if necessary.

PUT IT INTO PRACTICE

Key or code the plants located on your landscape plan. Record the codes on your informal list for future reference and use (refer to Exercise 25). The actual plant list will be drafted onto the plan sheet in the next exercise.

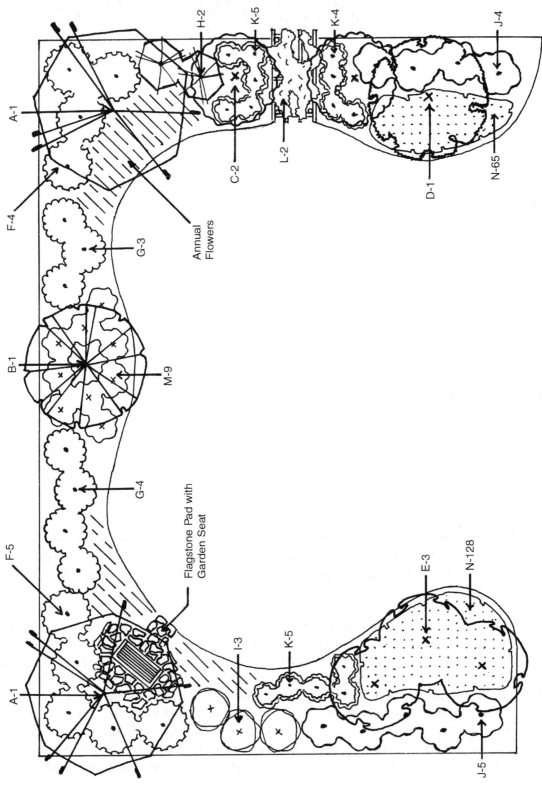

FIGURE 26–1 An example of keys or codes to identify plants on a landscape plan.

Exercise 26 Activity

Student Name _____ Date _____ Score _____

Evaluation

Consideration	Points	Student Score	Instructor Score
All plants, or groups, are coded	60		
Lettering is parallel to bottom edge	20		
Lines connect codes to plants	20		
Total	100		

NOTES

Preparing a Finished Plant List

OBJECTIVE

To assist students in drafting the final plant list onto the landscape plan.

SKILLS

After studying this unit, you should be able to:

- Draft the plant list onto the plan, providing appropriate information.
- Use guidelines to properly align codes, numbers, and names.
- Understand why botanical names are placed on the plan.

MATERIALS NEEDED

Drawing board and T-square

Drawing pencil (HB, F, H, or 2H)

Eraser and shield

Engineer's scale or architect's scale

Informal plant lists from Exercises 25 and 26

Drawing from Exercise 26

INTRODUCTION

An accurate plant list can provide a quick reference of the plants in the plan and the quantity of plants for each species. If accurate, the plant list alone can be used to purchase the plants for the landscape. You will need to provide four kinds of information for the plant list: codes, quantity of each plant, common names, and botanical names. Some plans go a step further and provide the size of the plant to be purchased.

1. Key—As previously discussed in Exercise 26, you will use letters of the alphabet to code the different plants.

2. Quantity (abbreviated Quan.)—Add quantities for each code, and list the total quantity under the heading Quan. For instance, you might have three of plant A in one area of the yard and five of plant A in another area. The quantity for plant code A would then be eight (3 + 5). See the sample sheet (Figure 27–1). Tally the total for each different code on your plan.

3. Common name—This is the English name used for a particular plant in your area, such as Sugar Maple or White Oak. Common names may vary from one area to another.

4. Botanical name—The botanical name, or scientific name for a plant, is the Latin name assigned to a plant. Two words are usually used. The first is the genus name, and the second is the species name. A third name is sometimes used, and it may be a variety name (Latin) or cultivar (any language and placed in single quotes). The botanical name is the only accurate name, because no two plants have the same exact botanical name.

In placing the information on the plant list, it will be necessary to use horizontal guidelines to provide the row of information for each plant. In addition, vertical guidelines will help align the information vertically in columns. Always leave space between the lines equal to or greater than the height of the letters used.

Study Figure 27–1 before beginning the activity.

PUT IT INTO PRACTICE

1. Using Figure 27–1 as a guide, prepare your plant list in the space reserved on your drawing. Use letter sizes that are approximately the sizes shown on the sample sheet, allowing slight variations to fit the scale you are using. Use both horizontal and vertical guidelines, as illustrated.

2. Be sure to include all plants on the design for the entire property.

Plant List

Key	Quan.	Common Name	Botanical Name
A	2	Red Maple	Acer Rubrum 'October Glory'
B	1	River Birch	Betula Nigra
C	2	Eastern Redbud	Cercis Canadensis
D	1	Yoshino Cherry	Prunus Yedoensis
E	3	Flowering Dogwood	Cornus Florida 'Cloud Nine'
F	9	Hybrid Rhododendron	Rhododendronx 'Roseum Elegans'
G	7	Manhattan Euonymus	Euonymus Kiautschovicus 'Manhattan'
H	2	English Holly	Ilex Aquifolium
I	3	Pampas Grass	Cortaderia Selloana
J	9	Mountain Laurel	Kalmia Latifolia
K	14	Kurume Azalea	Azalea obtusum 'Hino-Crimson'
L	2	Hybrid Clematis	Clematis X Jackmanii
M	9	Helleri Holly	Ilex Crenata 'Helleri'
N	193	Japanese Surge	Pachysandra Terminalis

FIGURE 27–1 Plant list.

Exercise 27 Activity

Student Name _____ Date _____ Score _____

Evaluation

Consideration	Points	Student Score	Instructor Score
List is aligned horizontally and vertically	30		
All plants are included	40		
Scale is accurate	20		
List is neat	10		
Total	100		

NOTES

Preparing a Title Block

OBJECTIVE

To guide students in the development of a symmetrical title block for the completed landscape plan.

SKILLS

After studying this unit, you should be able to:

- Draft a symmetrical title block for a residential landscape plan.
- Include all needed information in the title block.
- Center each line on a central, vertical guideline.

MATERIALS NEEDED

Drawing board and T-square

Drawing pencil (HB, F, H, or 2H)

Eraser and shield

Engineer's scale or architect's scale

Drawing from Exercise 27

INTRODUCTION

The title block is the final step in the preparation of the landscape plan. It is one of the first things that people notice, so it should be neat and should contain all necessary information as well as the words *Landscape Plan for* [the client's name], the month and year developed, the designer or designer's name, and the scale used. In some cases it is desirable to include the drawing number, client's address, designer's company or address, and the page number if more than one page is used.

In preparing the title block, the client's name should be drafted in larger letters than other lettering. After all, who's paying for the design? Designers might desire to emphasize their names, but their name or company name should be drafted in smaller letters than the client's name.

The title block should be centered in the remaining space on the lower right portion of the drawing. It is unnecessary to enclose the information in a block or box, as implied by the wording *title block*. However, some vellum comes with a true block, preprinted and allowing space for all necessary information. In fact, some businesses order custom-designed vellum that contains custom-made title blocks with preprinted company name or logo.

In the absence of preprinted title blocks, the most visually pleasing title block is the symmetrical title block. To prepare a symmetrical title block, it is necessary to use several horizontal guidelines and one vertical guideline. Basically, each horizontal line is centered on the vertical line (Figure 28–1).

One helpful technique is to draft the horizontal lines and wording on a separate sheet, slide the sheet under your drawing, center it on the vertical line, and trace it. However, it is possible to count the letters and spaces on a line and divide by two. This will help you to locate the center letter or space on a line. Remember: This is your last chance to make the design look really good. *Go for it!*

Study Figure 28–1, and use the letter sizes and spaces as indicated.

PUT IT INTO PRACTICE

1. Prepare a symmetrical title block on your plan. Make up your own client name, and list your name as the designer.

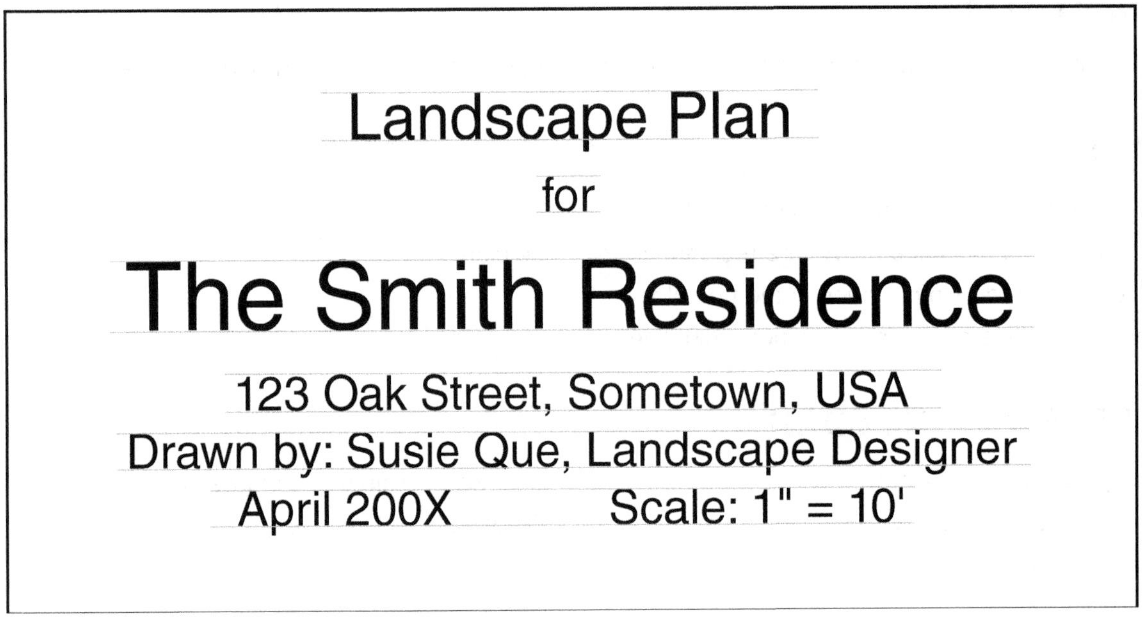

Landscape Plan
for
The Smith Residence
123 Oak Street, Sometown, USA
Drawn by: Susie Que, Landscape Designer
April 200X   Scale: 1" = 10'

FIGURE 28–1 Example of a symmetrical title block.

Exercise 28 Activity

Student Name _____ Date _____ Score _____

Evaluation

Consideration	Points	Student Score	Instructor Score
Light guidelines are used	30		
Title block is symmetrical	30		
Lettering is accurate and neat	40		
Total	100		

NOTES

Drafting Details on Landscape Plans

OBJECTIVE

To provide experience in drafting details for staking trees and constructing artifical features.

SKILLS

After studying this unit, you should be able to:

• Complete a scaled drawing for planting and staking landscape trees.

• Draft details for a artifical feature in the landscape.

• Write specific notes to accompany detail drawings.

MATERIALS NEEDED

Drawing board and T-square

Drawing pencil (HB, F, H, or 2H)

Eraser and shield

Engineer's scale or architect's scale

Completed landscape plan

Sample plans found in other exercises, if needed

Drafting vellum or bond copy paper

Drafting tape

INTRODUCTION

Drafting details on landscape plans (or on supplementary sheets) is not always necessary. Such details are necessary when there is a need for staking trees or when artifical features are included in the design to be constructed as part of a landscape package.

Staking trees is necessary for all trees planted bare root and should be maintained until the plants are well established, usually two years. Balled and burlapped trees (B&B) that have been dug by hand or harvested by machine and placed in burlap-lined wire baskets are easier to stabilize. However, these trees are usually staked when planted in open areas of lawn where windy conditions prevail and the trees are densely branched, making them more likely to tilt under extreme winds. Trees located in parking lots or bordering guest parking areas are subject to being jarred or bumped, and should be staked.

Indicating details for staking trees can usually be accomplished with a simple cross-sectional drawing to a scale large enough to clearly show what is to be done. Most details are drafted on a 1 × 4–inch (1:4) scale or even larger, such as 1 × 2–inch scale (1:2). Smaller scales are acceptable so long as detail is clearly visible. Many designers draft such details once and file them for future use on other designs.

Figure 29–1 shows a staking detail with labels and instructional notes.

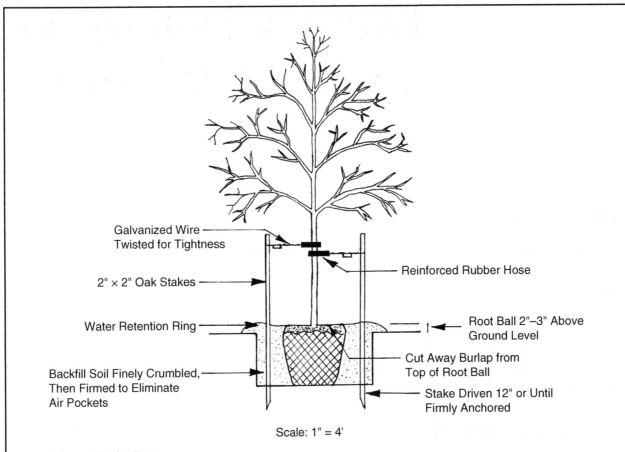

Galvanized Wire Twisted for Tightness

2" × 2" Oak Stakes

Water Retention Ring

Backfill Soil Finely Crumbled, Then Firmed to Eliminate Air Pockets

Reinforced Rubber Hose

Root Ball 2"–3" Above Ground Level

Cut Away Burlap from Top of Root Ball

Stake Driven 12" or Until Firmly Anchored

Scale: 1" = 4'

Notes for Staking Trees

1. Planting hole should be twice as wide as root ball or a minimum clearance of 8″ to 10″ in any direction.

2. Oak stakes should be long enough after anchoring to allow placement of wire just below the lowest branches of the tree.

3. Remove top of burlap and any twine from lower trunk of tree.

4. Soft galvanized steel wire should be 14 gauge or stronger.

5. Wire should loop around stake and through rubber hose. After connecting the ends, place 1" × 2" board between and twist to tighten.

6. Planting tree slightly above surrounding ground prevents drowning. Use 2" bark mulch or pine straw mulch inside water ring.

FIGURE 29–1 A staking detail with labels and instructional notes.

Details for artifical features are numerous and sometime require two or more drawings. Examples of some commonly detailed features include patios or walks set in sand, trellises and arbors, garden ponds, berms (mounds of soil with turf, ground cover, shrubs, or trees), raised planters, lawn edging or mowing strips, fences, and garden benches.

NOTE: Designers often design the size, shape, and materials for surfaces and walls built with brick and mortar, decks, playhouses, and storage buildings. These items often require more expertise in construction detail and should be constructed by skilled professionals in accordance with local building codes.

Figure 29–2 shows detailed drawings for a brick entrance walk.

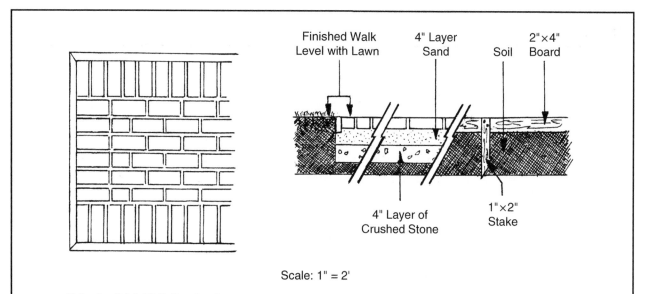

Finished Walk Level with Lawn   4" Layer Sand   Soil   2"×4" Board

4" Layer of Crushed Stone   1"×2" Stake

Scale: 1" = 2'

Notes for Brick Walk Construction

1. Lay out walk area accurately and mark with "upside down" paint or lime. Dig to a depth of 8".

2. Evenly spread 4" nut-size crushed stone.

3. Install 2" × 4" board edging (pressure treated against decay). Attach 1" × 2" treated stakes to outside of 2" × 4" boards with galvanized wood screws (2"). Length of stakes to be 12" or more, as needed for firm anchoring.

4. Apply 4" sand layer. Level and firm.

5. Lay brick in pattern shown, leaving 5/8" spacing on all sides. Tap with rubber mallet until firmly seated at ground (lawn) level.

6. After all bricks are in place, sweep sand/cement mixture over bricks to fill gaps. Spray water over bricks. Fill any gaps left with the mixture, then water again. Allow 2 days before using walk.

FIGURE 29–2 Detailed drawings for a brick entrance walk.

PUT IT INTO PRACTICE

Use your completed landscape plan for this activity. If your plan does not contain any features that can be detailed, you may use a feature from any of the sample plans in previous exercises.

1. Draw an arrow to the feature on the plan and label it with an Ⓐ See detail, Ⓑ See detail, and so on. Each feature that has a detail should be labeled with a letter.

   a. On a separate sheet (vellum, graph paper, or plain printer/copier paper) draw a construction detail, using one or more views to show precise detail.

   b. Add any notes that are necessary for exact construction.

2. Label the scale you use for the detail.

NOTES

Exercise 29 Activity

Student Name _____ Date _____ Score _____

Evaluation

Consideration	Points	Student Score	Instructor Score
Drawing scale is of adequate size	20		
Notes are adequate in detail	30		
Labels and notes are neat	20		
Location of detail on plan is clearly identified	30		
Total	100		

NOTES

SECTION TWO

Landscape Calculations

Creating a Cost Estimate of the Landscape Design Plan

OBJECTIVE

To provide experience in estimating the cost of a landscape design.

SKILLS

After studying this unit, you should be able to:

- Complete an estimate of what a landscape plan will cost.

MATERIALS NEEDED

Calculator (*optional*)

INTRODUCTION

A large amount of time is spent on developing a plan for the landscape and the customer. However, if the landscape plan will cost more than the customer has budgeted, then the hard work that was invested can easily become useless. It is important that landscape architects and designers take budget into account when designing. It is also important for them to have a strong knowledge of how to create an estimate of the landscape plan. An estimate is an approximation of the price that a customer will pay for a landscape plan. It is important that when the estimate is presented to a client, he or she understands that the amount is an approximation and could be lower or higher, depending on certain situations.

Preparing the Estimate

Landscape estimates and bids are prepared as spreadsheets. As the data are assembled, they are presented in columns with itemized quantities, descriptions of materials or services, unit costs of materials, unit costs of installation, and total costs for each item. The data are also grouped into categories organizing the estimate for easy reading and referral (Figure 30-1). A typical design cost estimate includes the following:

- Cost of site clearing and other preparation
- Cost of plant materials
- Cost of construction materials
- Cost of turfgrass
- Allowance for overhead
- Allowance for contingencies
- Fee for landscape designing
- Allowance for profit
- Name of the estimator and date of the estimate

The fee for landscape design services is handled in several ways, depending on the size of the project, the credentials of the designer, the policy of the firm, and the laws of the state. In some states, only accredited landscape architects are permitted to charge for their design services. Design-build firms often have salaried designers on the payroll. In such companies, the design costs are usually treated as overhead costs assigned to specific projects. In either case, the fee for designing, drafting, and overseeing the installation of a project to meet full client satisfaction can be estimated at 8 to 15 percent of the total cost of materials and installation. A large project usually uses the lower percentage of the range, and a small project uses the higher percentage.

Price Estimate for the Design and Development of the Property of Mr. and Mrs. John Doe, 1234 Main Street, Cleveland, Ohio

I. PLANT MATERIALS

Description	Quantity	Unit	Material Cost	Installation Cost	Total + 35%
Celtis occidentalis, 1 1/2" cal. B&B	6	EA	$125.00	$31.00	$1,263.60
Cornus florida, 7', B&B	5	EA	101.00	45.00	985.50
Euonymus alatus, 15", BR	20	EA	7.00	10.55	473.85
Philadelphus coronarius, 3', BR	15	EA	5.40	15.63	425.86
Vinca minor, 2" pots	50	EA	1.05	0.50	104.63
			Total for plant materials		$3,253.44

II. CONSTRUCTION MATERIALS

Description	Quantity	Unit	Material Cost	Installation Cost	Total + 25%
Concrete pavers, $3^{1}/_{8}$" thick, interlocking dry joints on 2" sand base with 4" gravel subbase	500	SF	$2.48	$1.53	$2,506.25
Pressure treated timber wall, 6" × 6" timbers, gravity type, inc. excavation and backfill	300	SF	6.00	7.00	4,875.00
Flagpole, aluminum, 25' ht. cone, tapered with hardware halyard and ground sleeve	1	EA	1,060.00	59.02	1,398.78
			Total for construction materials		$8,780.03

III. TURFGRASS

Description	Quantity	Unit	Material Cost	Installation Cost	Total + 35%
Kentucky bluegrass sod on level prepared ground, rolled and watered	900	SY	$1.00	$1.48	$3,013.20
			Total for turfgrass		$3,013.20

IV. TOTAL COST OF ALL MATERIALS AND INSTALLATION (See Note 1) — 15,046.67

V. CONTINGENCY ALLOWANCE (See Note 2) — 1,504.67

VI. FEE FOR LANDSCAPE DESIGN SERVICES (See Note 3) — 1,805.60

VII. TOTAL COST FOR COMPLETE LANDSCAPE DEVELOPMENT (See Note 4) — $18,356.94

FIGURE 30–1 An example of a completed cost estimate.

NOTES: (1) The total cost is obtained by adding the subcategory totals.

(2) The contingency is taken as a percentage of total costs in IV. In the example, 10% is used.

(3) The design fee is taken as a percentage of IV also. In the example, 12% is used.

(4) The final figure should be the most distinctive on the spreadsheet. It should be the bottom line figure, with no others below it to create confusion.

PUT IT INTO PRACTICE

Prepare a complete price estimate for the property of Mr. and Mrs. Chad Hamm, 123 Main Street, Nicholasville, Kentucky. Use the data in Exercise 30 Activity and follow the format outlines in this exercise. No site preparation is required. Optional: Exercise could be completed using spreadsheet software if available.

NOTES

Exercise 30 Activity

Student Name _____ Date _____ Score _____

PLANT MATERIALS REQUIRED

2   Chinese redbud (*Cercis chinensis*), 5' ht., B&B, cost $59.00 each and $16.25 to install

20   Golden forsythia (*Forsythia spectabilis*), 18" ht., BR, cost $3.00 each and $11.63 to install

2   Sugar maple (*Acer saccharum*), 1 1/4" cal, B&B, cost $58.00 each and $30.85 to install

10   Japanese barberry (*Berberis thunbergi*), 15" ht., CT, cost $19.50 each and $10.55 to install

20   Vanhoutte spirea (*Spiraea vanhouttei*), 1 gal. CT, cost $6.75 each and $10.55 to install

300   Japanese pachysandra (*Pachysandra terminalis*), 2" pot, cost $0.83 each and $0.50 to install

1   Babylon willow (*Salix babylonica*), 8' ht., B&B, cost $44.70 and $45.00 to install

12   Floribunda roses (*Rosa floribunda*), 18" ht., BR, cost $6.60 each and $11.63 to install

2   Red oak (*Quercus rubra*), 6' ht., BR, cost $35.75 each and $23.48 to install

---

NOTE: Profit and overhead allowance: 35% of the cost of materials and installation.

CONSTRUCTION REQUIRED

9   gauge fabric @ $239.00 each; installation cost is $8.49 each

225   square feet of brick pavers, 2 1/4" thick, set over a finished subgrade, dry joints, 2" sand base @ $2.88 per SF; installation cost is $2.71 per SF

4   walk lights, 10' high, plain steel pole set into a cubic foot of concrete, with distribution system and control switch @ $1875.00 each; installation cost is $350.00 each

---

NOTE: Profit and overhead allowance: 25% of the cost of materials and installation.

LAWN REQUIRED

1,000   square yards of Kentucky bluegrass sod, rolled and watered @ $3.50 per SY. Installation cost is $0.72 per SY.

NOTE: Profit and overhead allowance: 35% of the cost of materials and installation.

Fee for landscape design services is to be 16% of the total cost of all materials and installation. Contingencies will be included as 10% of the total cost of all materials and installation. Use your name and the current date as the name of the estimator and the date of the estimate.

NOTES

Pricing Landscape Maintenance

OBJECTIVE

To provide experience in developing cost analysis for common landscape maintenance jobs.

SKILLS

After studying this unit, you should be able to:

- Complete an accurate analysis of the cost of specific landscape maintenance tasks.

MATERIALS NEEDED

Calculator (*optional*)

INTRODUCTION

As young landscapers begin their business, they often don't take the time that is necessary to develop accurate cost analysis for jobs they plan to perform. To ensure success as well as provide a fair price to your customer, it is important for these steps to be taken.

There are two ways many landscapers determine cost for maintaining properties. The first one is **comparison pricing**. This procedure is predominately used by established companies that have time-tested data that allow them to predict cost of completing jobs. This is not a preferable method for a young company. The second way, which is more favorable and more precise, is **unit pricing**. Unit pricing does take more time and effort to set up, but the overall end result is more accurate. In order to provide unit pricing, the following items must be included:

- A list of all tasks to be performed
- The total square footage area involved for each service
- The number of times each service is performed during the year
- The time required to complete each task once
- The time required to complete each task annually
- The cost of all materials required for each task
- The cost of all labor required for each task

To apply the technique properly requires practice. Study the following examples and their explanations.

EXAMPLE 31-1

Problem
To calculate the cost of mowing 20,000 square feet of lawn with a 52-inch zero-turn mower 25 times each year.

Landscape Information
It takes 3 minutes to mow 5,000 square feet. The laborer receives $11.00 per hour. There are no material costs.

Solution
Maintenance operation: Lawn mowing with 52-inch zero turn.

Square footage area involved	20,000
Number of times performed annually	25
Minutes per 5,000 sq. ft.	3
Total annual time in minutes (1)	300
Material cost per 1,000 sq. ft.	none
Total material cost	none
Wage rate per hour	$11.00
Total labor cost (2)	$55.00
Total cost of maintenance operation per year	$55.00

NOTES: (1) To obtain the total annual time in minutes:

    a. Divide the square footage of area involved by 5,000 (20,000 sq. ft. ÷ 5,000 = 4).

    b. Multiply by minutes per 5,000 sq. ft. (3 × 4 = 12 minutes).

    c. Multiply by number of times performed annually (12 minutes × 25 = 300 minutes).

    d. Enter answer under total annual time in minutes.

(2) To obtain the total labor cost:

    a. Divide the total annual time in minutes by 60 minutes (300 minutes ÷ 60 minutes = 5 hours).

    b. Multiply by the wage rate per hour (5 hours × $11.00 = $55.00).

    c. Enter answer under total labor cost.

EXAMPLE 31-2

Problem

To calculate the cost of mulching 2,000 square feet of planting beds with wood chips, 4 inches deep.

Landscape Information

The task is done once each year. It requires 30 minutes to mulch 1,000 square feet, 4 inches deep. The laborer receives $11.00 per hour. The wood chips cost $165.00 per 1,000 square feet of coverage.

Solution

Maintenance operation: Mulching plantings with wood chips, 4 inches deep.

Square footage area involved	2,000
Number of times performed annually	1
Minutes per 1,000 sq. ft.	30
Total annual time in minutes (1)	60
Material cost per 1,000 sq. ft.	$165.00
Total material cost (2)	$330.00
Wage rate per hour	$11.00
Total labor cost (3)	$11.00
Total cost of maintenance operation per year (4)	$341.00

NOTES: (1) To obtain the total annual time in minutes:

    a. Divide the square footage of area involved by 1,000 (2,000 sq. ft. ÷ 1,000 = 2).

    b. Multiply by minutes per 1,000 sq. ft. (2 × 30 = 60 minutes).

    c. Multiply by number of times performed annually (60 minutes × 1 = 60 minutes).

    d. Enter answer under total annual time in minutes.

(2) To obtain the total material cost:

    a. Divide the square footage of area involved by 1,000 (2,000 sq. ft. ÷ 1,000 = 2).

    b. Multiply by material cost per 1,000 sq. ft. (2 × $165.00 = $330.00).

    c. Multiply by number of times performed annually ($330.00 × 1 = $330.00).

    d. Enter answer under total material cost.

(3) To obtain the total labor cost:

    a. Divide the total annual time in minutes by 60 minutes (60 minutes ÷ 60 minutes = 1 hour).

    b. Multiply by the wage rate per hour (1 hour × $11.00 = $11.00).

    c. Enter answer under total labor cost.

(4) To obtain the total cost of maintenance operation per year:

    a. Add total material cost and total labor cost ($330.00 + $11.00 = $341.00).

    b. Enter answer under total cost of maintenance operation per year.

PUT IT INTO PRACTICE

Complete the following landscape maintenance pricing problems. Use the chart to assist you in organizing your data.

NOTES

Exercise 31 Activity I

Student Name _____ Date _____ Score _____

Landscape Information

The task is done twice a year. It requires 30 minutes to prune 1,000 square feet of barberry hedge in front of the nursing home on Main Street. The total amount of barberry hedge to prune is 9,500 square feet. The laborer receives $15.00 per hour.

Solution

Square footage area involved	
Number of times performed annually	
Minutes per 1,000 sq. ft.	
Total annual time in minutes	
Material cost per 1,000 sq. ft.	
Total material cost	
Wage rate per hour	
Total labor cost	
Total cost of maintenance operation per year	

Exercise 31 Activity II

Student Name _____ Date _____ Score _____

Landscape Information

The task is done once each year. It requires 30 minutes to mulch 1,000 square feet, 4 inches deep. The area to be covered with mulch is 3,500 square feet. The laborer receives $12.50 per hour. The black mulch cost $180.00 per 1,000 square feet of coverage.

Solution

Square footage area involved	
Number of times performed annually	
Minutes per 1,000 sq. ft.	
Total annual time in minutes	
Material cost per 1,000 sq. ft.	
Total material cost	
Wage rate per hour	
Total labor cost	
Total cost of maintenance operation per year	

SECTION THREE

Customer Service

Communication and Marketing of Landscape Business

OBJECTIVE

To provide experience in developing strategies to communicate and market your business to customers.

SKILLS

After studying this unit, you should be able to:

- Develop communication pieces for customers using different types of strategies.

MATERIALS NEEDED

Digital devices (computer, tablet, smartphone)

Internet connection

INTRODUCTION

Communication with potential customers is key to gaining new business as well as maintaining existing customers. Over the years, communication and marketing of landscaping businesses has drastically changed. A few past marketing strategies would include being listed in the yellow pages of the phone book, sending out letters to potential customers, or putting out signs in highly traveled areas. Although some strategies from the past can still be used today, they should not remain as your only means of communicating with customers.

Email, smartphones, and tablets have revolutionized how you communicate with customers and market your business and services. A few ways that landscaping businesses find beneficial are listed here:

- Social media business pages (Facebook, Instagram, Twitter, etc.)
- Email to potential customers
- Digital coupons

Social media business pages are a great way to showcase your company. Potential customers and existing customers can follow your pages to see your current projects, see specials you may be running, and keep up with your business. By planning strategic posts, you can keep your business in front of customers all hours of the day. It is important to remember that social media should be used in a manner in which both parties feel engaged. Provide customers advice for keeping up their landscaping, or provide customers with an opportunity to save money by following your page. There are so many ways to engage customers, from having contests to encouraging feedback on the social media forum. As social media and the digital platforms continue to evolve, you should find new ways to market your business.

PUT IT INTO PRACTICE

1. Create a Twitter post using no more than 140 characters to advertise a special your business is having on spring mulching. You decide what type of discount you plan to give. Don't forget to create your Twitter handle for your business. Your Twitter handle should begin with @, for example, "@lawncare4u."

2. Develop a series of three social media posts that would be posted during the fall months to help homeowners with fall pruning and leaf removal. These should be posts that are helpful to homeowners to grow your customer base. If possible, include pictures and examples for homeowners. Remember that even though word count isn't an issue, short and concise posts provide better readability. The use of #hashtags would be valuable as well.

NOTES

Exercise 32 Activity

Student Name _____ Date _____ Score _____

Evaluation

Consideration	Points	Student Score	Instructor Score
Creativity and originality	30		
Accurate information regarding landscaping	40		
Followed directions & included all components	30		
Total	100		

NOTES

Measuring Customer Satisfaction

OBJECTIVE

To provide experience in developing strategic surveys that allow for your business to grow because of customer satisfaction.

SKILLS

After studying this unit, you should be able to:

• Develop questions that will allow you to measure your customer satisfaction level.

MATERIALS NEEDED

Computer access

Online access (to research survey examples)

INTRODUCTION

As you develop your business, you will find that surpassing customer expectations and improving satisfaction will increase loyalty. If customers are happy and enjoyed their experience with your company, they will be more apt to return in the future, as well as recommend your services to others. Understanding what makes your customers happy is important to growing your business but is often overlooked as an essential part.

A great way to gather information from customers is through customer surveys. Customer surveys can be simple and easy to fill out on paper, online, or through a phone call. Providing the opportunity for customers to give feedback allows for all customers to feel as if they are a part of your team, and it provides you the opportunity to reflect on what you are doing well and what you can improve. No matter how you distribute your survey, there are some important questions that need to be included:

• Speed of service – Consider the three stages of service: before the purchase, during the purchase, and after the purchase. For landscaping jobs, this might read before the contact, during the job, and after the job completion.

• Quality of service – Consider not only the quality of the job or service you performed but also attitudes of staff, knowledge of staff, and responsiveness to inquiries or complaints.

- Pricing – Consider not only the total cost of your service or job completed but also include the value for the money.
- Overall satisfaction with service or job – Overall, how satisfied are they with the job or service performed?
- Customer loyalty – How likely are they to recommend your company to another individual?
- Intention to repurchase – Are they likely to repurchase this service or product within the next six months or year or purchase another product or service you supply?
- Other needed services – Ask if there is anything else they may need from your company.

Other questions you might ask include where they may have heard about your company. This allows you to understand what avenue of marketing is working best for you.

Developing the survey to allow for easy understanding of the data is important. Although open-ended questions seem like the easy approach, they are very hard to compile and desegregate the data. Opt for closed-ended questions and those that can be evaluated with a rating scale. Closed-ended questions may be "Will you purchase a service from our company again in the future?" That simple yes-or-no question will give you a clear indication of the customer's future intent. For other questions, you may use a 1 to 5 rating scale, where 1 is extremely likely and 5 is not likely. For example "How likely are you to purchase a service from our company again in the future?" With this type of question, you will find the data are more objective and easier to measure. You can combine all scores to find a mean score. A mean score that is closer to 1 in this example is very good and means the person plans to purchase in the future. A mean score closer to 5 shows the person doesn't plan to purchase in the future.

Creating a survey that is easy for customers to understand and take should be your goal. You also want it to be easy to compile the data so you can easily understand what has been shared. Once you compile your data, take time to reflect on your successes and how you can improve in the areas that may need improvement.

PUT IT INTO PRACTICE

Create a 10-question survey to evaluate your mowing service. Use closed-ended questions and a rating scale. Include one open-ended response question. Use the information in the Introduction to help you formulate questions. Review the Evaluation rubric before beginning.

NOTES

Exercise 33 Activity

Student Name _____ Date _____ Score _____

Evaluation

Consideration	Points	Student Score	Instructor Score
Included 10 questions (closed-ended and rating scale questions with one open-ended question)	30		
Questions will accurately define customer service (do they follow the important questions provided in the Introduction?)	40		
Questions are easy to understand	30		
Total	100		

NOTES

Handling Customer Complaints

OBJECTIVE

To provide experience in handling customer complaints in a professional manner.

SKILLS

After studying this unit, you should be able to:

- Respond to customer complaints in a professional manner.

MATERIALS NEEDED

Pen/pencil

INTRODUCTION

The landscaping industry is a service industry. Customers expect a certain level of service from the company they do business with. The way businesses receive complaints has changed over the years with the digital age. Previously, a complaint might have been brought in to the business in person or received over the phone. It might also have come in the mail in the form of a letter. Today, many complaints are digital in the form of emails, survey forms, or posts on social media platforms like Facebook or Twitter. No matter what form they come in, it is important that they are attended to quickly and in a professional manner. Here are some strategies for handling customer complaints.

- Stay calm and listen to the customer. Do not interrupt; let them finish!
- Acknowledge the problem. Ask questions in a caring manner, and get the details needed.
- Do not blame but apologize for the problem.
- Solve the problem or find someone who can as quickly as possible. Do not delay.

When dealing with customer complaints, it is important not to argue, but to acknowledge their concern. By arguing with a customer, you are only sure to escalate the problem even further. A question you might ask the customer is "What would be an acceptable solution for you?" This gives the customer the opportunity to partner with you on finding a solution. It is always good to be prepared to offer some type of solution to the problem when you ask this question. The last thing you want to do is promise something you cannot make good on. It would be better to say, "Let's see if we can find a solution that will work for both of us," or "We will check to see how we can make this situation better." The most important thing is to follow up on your promise and get back to the customer quickly.

PUT IT INTO PRACTICE

Please read the customer complaint email in Figure 34-1, and respond in an appropriate manner in the blank email that follows. Review the Evaluation rubric before beginning your response.

NOTES

Exercise 34 Activity

Student Name _____ Date _____ Score _____

To: lawnescapes@lawns.com
From: jill.england@qrx.com
Subject: Issues with trees

My name is Jill England, and I had you trim the burning bushes at my rental house. Since that time, I have noticed one bush is not doing very well. The leaves are turning brown and dead looking, and I am not sure what is going on. There is also one place where they have completely died and the branches are bare. I am very disappointed because I have had these bushes as a part of my landscape for several years. I am not sure I would recommend your company to anyone since this has happened.

I appreciate your time and look forward to hearing from you regarding a solution.

Sincerely,
Jill

To: jill.england@qrx.com
From: lawnescapes@lawns.com

RE: Issues with trees

_____
_____
_____
_____
_____
_____
_____
_____
_____
_____
_____
_____
_____
_____
_____

Sincerely,

_____ [Name]
_____ [Title]
Lawn Escapes

FIGURE 34–1 Customer complaint email.

Evaluation

Consideration	Points	Student Score	Instructor Score
Acknowledged the complaint	25		
Asked for more details to assist in solving problem	25		
Provided options to solve the problem	25		
Provided a professional response	25		
Total	100		

SECTION FOUR

Landscape Maintenance

Operating a Lawn Mower

OBJECTIVE

To learn the proper operation of a lawn mower.

SKILLS

After studying this unit, you should be able to:

* Operate a lawn mower in a proper and safe manner.

MATERIALS NEEDED

Push rotary mower

Zero-turn mower

Lawn area to practice mowing techniques

INTRODUCTION

Operation of a lawn mower can be new and exciting when you are first learning how to use it. However, the lawn mower is a piece of heavy equipment that should be used with precaution. Before operating the lawn mower, it is very important to reference the owner's manual. Each brand of lawn mower is made differently, and it is important that you pay special attention to the safety section of the owner's manual as well as information related to its operation, which will give you specifics on how to safely use this piece of equipment.

PUT IT INTO PRACTICE

Your instructor will demonstrate safe operation of equipment and administer both an oral and written exam based on the demonstration. You must score 100 percent in order to progress to operation of the equipment. Practice the correct mowing techniques as approved by your instructor. Please review the Evaluation rubric before beginning.

1. Do not operate the mower until you have read the entire manufacturer's manual.

2. Conduct a safety and engine check before operating:
   a. Make sure all the guards are in place before starting the machine.
   b. Check all shields and tension of the belts before mowing.
   c. Check engine fluids—fuel (gas or diesel) and engine oil—and lubricate all grease fittings according to the recommendations of the owner's manual.
   d. Check tire pressure and add air if needed.
   e. Before mowing, check the area you are mowing for objects lying in the lawn. Be sure all is clear before mowing.

3. Motor fuels are extremely dangerous and flammable. Do not permit open flames or sparks near the engine at any time.

4. Never stand in front of the mower while it is running.

5. Do not adjust the mower or perform any maintenance functions while the engine is running.

6. Never run the engine in an area without ventilation.

Checking Tire Pressure on Your Lawn Mower

The following are recommended steps for checking the tire pressure in your lawn mower:

- Check tire pressure when the tires are cold (the mower has been parked for 3 hours or more).
- Check your owner's manual to find the correct tire pressure or check the sidewall of the tire. The maximum tire pressure is displayed there. The factory settings for residential riding mowers are usually 14 pounds per square inch (psi) for the front tires and 10 psi for the rear tires.
- You will measure tire pressure in psi using a tire pressure gauge (Figure 35-1).
- To check the tire pressure, remove the tire valve stem cap, and press the tire pressure gauge on the tire valve stem. The gauge provides a digital reading (if a digital gauge) or displays a number on the meter stick.
- Compare your reading to the maximum tire pressure found on the wall of the tire to decide whether the tire is overinflated or underinflated.

FIGURE 35-1 A tire gauge.

Engine Service Maintenance

Every 25 hours:	**Servicing Air Filter:**
	Service the foam air filter by washing in soap and water, rinse, and dry. Re-oil with clean engine oil.
	Change the oil (consult manufacturer's manual). Change the oil more frequently under dusty conditions.
Every 50 hours:	Service the fuel filter.
Every 100 hours:	**Servicing Air Filter:**
	Service the air filter cleaner paper element by gently tapping on the flat side. Do not wash or use pressurized air. Replace the paper element each year or more often under dusty conditions.

Operation of Mower

1. When using a push mower, never pull it back toward your feet. If you have to back up, hold the handle with your right hand, turn around, and walk forward while pulling the mower back.

2. Now mow two widths of the mower deck around the boundary of the property area.

3. Mow in a north–south direction, and rotate direction 15 degrees with each consecutive manicure of the lawn area.

4. If there are edged and mulched beds in the lawn area, *always* mow with the left side of deck close to the mulched bed, and this will prevent discharging grass clippings into the mulched bed.

5. Follow this procedure until the lawn area has been completely manicured.

6. This process will result in a beautiful lawn.

NOTES

Exercise 35 Activity

Student Name _____ Date _____ Score _____

Evaluation

Consideration	Points	Student Score	Instructor Score
Read owner's manual prior to operation	25		
Performed safety and engine inspection prior to mowing	25		
Mowed correctly based on direction and order of mowing	25		
Followed safety precautions while mowing	25		
Total	100		

NOTES

Operating a Commercial Dethatcher

OBJECTIVE

To operate a commercial dethatcher.

SKILLS

After studying this unit, you should be able to:

- Operate a commercial dethatcher in a proper and safe manner.

MATERIALS NEEDED

Commercial dethatcher

Lawn practice area

INTRODUCTION

Thatch is the layer between the belowground root system of turf and the aboveground living turf. Thatch is normally comprised of stems, roots, and leaves. As long as the thatch stays less than a half-inch thick, it can actually help to mulch the soil and insulate the belowground root systems, as well as having other benefits. But once it gets thicker than half to three-quarters of an inch, then it is necessary to dethatch your lawn. Large lawns that need dethatching are best completed by using a commercial dethatcher. Dethatchers are large equipment that should be used with precaution. If you are renting or own the equipment, make sure to read your owner's manual to understand the features and safety issues that may arise when using a dethatcher.

PUT IT INTO PRACTICE

CAUTION

Use extreme care when handling a commercial dethatcher!
It should be operated by trained
individuals only!

1. Read the following information.

2. Your instructor will demonstrate how to properly and safely operate a commercial dethatcher.

3. Practice the operation of a dethatcher as approved by the instructor. Please review the Evaluation rubric before beginning.

SAFETY PRECAUTIONS

1. Read the manufaturer's manual thoroughly before operating or maintaining the dethatcher.

2. Always wear safety glasses, long pants, and safety shoes when operating or maintaining this machine. Do not wear loose-fitting clothing.

3. Always disconnect the spark plug wire before adjusting or maintaining this machine.

4. Never put your hands or feet under the deck assembly unless for maintenance, with the spark plug wire disconnected.

5. Do not remove any shields, guards, or decals. If a shield, guard, or decal is damaged or does not function, repair or replace it before operating the dethatcher.

6. Altering this machine in any way may cause injury to the operator or bystanders.

7. Make certain the dethatcher is in the transport position when starting the engine.

8. When moving the dethatcher, always place the unit in the transport position, and push the unit from behind.

9. Keep people and pets away from the dethatcher when in use.

10. Remove all foreign objects in the path of the dethatcher.

11. Watch out for sidewalks, curbs, rocks, small tree stumps, roots, sprinkler heads, stakes, water shutoff plates, and so on.

12. If a breakdown occurs, shut off the engine immediately, disconnect the spark plug wire, and do not reconnect until repairs are made.

13. Never run the engine indoors without adequate ventilation. Exhaust fumes are deadly.

14. Gasoline is extremely dangerous and flammable. Do not permit open flames or sparks near the engine at any time.

15. Keep the dethatcher, and especially the engine and belt area, clean and free of grease, grass, and leaves, to reduce the chance of fire and to permit proper cooling.

Engine Service Maintenance

Every 25 hours:	**Servicing Air Filter:**
	Service the foam air filter by washing in soap and water, rinse, and dry. Re-oil with clean engine oil.
	Change the oil (check manufacturer's manual). Change the oil more frequently under dusty conditions.
Every 50 hours:	Service the fuel filter.
Every 100 hours:	**Servicing Air Filter:**
	Service the air filter cleaner paper element by gently tapping on the flat side. Do not wash or use pressurized air. Replace the paper element each year or more often under dusty conditions.

Operating Instructions Prior to Each Day's Use

1. Disconnect the spark plug wire.
2. Position the dethatcher on a level surface and check the engine oil and gear reduction oil levels. Add fluids as necessary.
3. Check the belt tension.
4. Check the condition of the blades. Adjust the depth control cam if necessary.
5. Check and tighten all nuts and bolts as necessary.

To Start the Dethatcher

1. Pull the control handle to the rear to lift the deck assembly to the transport position, and push the dethatcher to the area to be dethatched.
2. Connect the spark plug wire.
3. Move the engine choke from "Run" to "Choke." Move the engine throttle from "Off" to "Idle."
4. Grasp the starter grip, and pull the cord slowly until the starter engages; then pull rapidly to overcome compression, prevent kickback, and start the engine. Allow the cord to recoil slowly. Repeat if necessary.
5. Move the choke back to "Run" and after the engine has warmed up, move the throttle to "Fast."
6. Stand behind the dethatcher, and move the machine to the point where you want to begin dethatching. Then push the control handle forward to lower the deck assembly so that the blades are in the dethatching position. Move slowly forward and begin dethatching. To make a turn, push down on the upper handle, and pivot the dethatcher on its rear wheels, with the blades disengaged from the turf. This will help to prevent bending or breaking the blades.
7. To shut off the machine, pull the control handle to the rear to lift the deck assembly to the transport position, move the engine throttle to "Off," and disconnect the spark plug wire.
8. After use, thoroughly clean the machine, especially the engine, the belt area, and under the cutter deck.

NOTES

Exercise 36 Activity

Student Name _____ Date _____ Score _____

Evaluation

Consideration	Points	Student Score	Instructor Score
Read owner's manual prior to operation	25		
Performed safety and engine inspection prior to dethatching	25		
Dethatched lawn according to instructions provided	25		
Followed safety precautions while dethatching	25		
Total	100		

NOTES

Operating a Lawn Aerator

OBJECTIVE

To safely operate an aerator to improve the aeration of the lawn.

SKILLS

After studying this unit, you should be able to:

- Operate a lawn aerator in a proper and safe manner.

MATERIALS NEEDED

Lawn aerator

Lawn area to practice

INTRODUCTION

Soil compaction in lawns can become a big problem that hinders lawn quality. To combat soil compaction, it is important to aerate your lawn, using a lawn aerator. Aeration of the lawn involves perforating the soil with small holes or removing a core of grass from the soil to allow water, air, and nutrients to penetrate the roots. This helps to produce a stronger, deeper root system that will become heartier. Lawn aerators that remove plugs of soil are often commercial equipment that should be used with precaution. Make sure you read the owner's manual before using the equipment. It is important to understand the safety precautions as well as other features of the equipment before operating it.

PUT IT INTO PRACTICE

CAUTION

Use extreme care when handling the aerator! It should be operated by trained individuals only!

1. Read the following information.
2. Your instructor will demonstrate how to properly and safely operate the lawn aerator.
3. Practice the correct aeration techniques as approved by the instructor. Please review the Evaluation rubric before beginning.

Safety Instructions

1. The operator should have access to the owner's manual at all times.
2. Do not operate this machine until you have read this entire manual thoroughly.
3. Make sure all guards are in place before starting this machine.
4. Never put hands or feet under the machine housing, except for maintenance. Do not adjust the machine or perform any maintenance functions while the engine is running. Always use handles when lifting.
5. Make certain the belt tensioner is not engaged when starting the engine.
6. Never stand in front of the machine while the engine is running.
7. CAUTION: The machine may start suddenly when the belt tensioner is engaged. Start machine at low throttle, and gradually increase to desired operating speed.
8. If a breakdown occurs, shut off the machine immediately, and do not restart until repairs are made.
9. Never run the engine in an area without proper ventilation.
10. Gasoline is extremely dangerous and flammable. Do not permit open flames or sparks near the engine at any time.
11. Remove all foreign objects in the path of the aerator.
12. Do not run the aerator under low-hanging limbs that will interfere with the operator.
13. When transporting the unit, always push the unit from behind.
14. When parking the aerator on a slope, shut off the engine, lower the tines, and engage the belt tensioner. Never leave the aerator on a slope in the transport mode.
15. CAUTION: Do not exceed manufacturer's recommended tire pressure. Failure to comply may cause tire to explode.
16. Altering this machine in any way may cause injury to the operator or bystanders.

Engine Service Maintenance

Every 25 hours:	**Servicing Air Filter:**
	Service the foam air filter by washing in soap and water, and then rinse and dry. Re-oil with clean engine oil.
	Change the oil (check manufacturer's manual). Change the oil more frequently under dusty conditions.
Every 50 hours:	Service the fuel filter.
Every 100 hours:	**Servicing Air Filter:**
	Service the air filter cleaner paper element by gently tapping on the flat side. Do not wash or use pressurized air. Replace the paper element each year, or more often under dusty conditions.

OPERATING INSTRUCTIONS

Prior to Each Day's Use

1. Check engine oil and gear reduction oil levels.
2. Lubricate all grease fittings (four wheels, two caster wheel yokes, spoon disc and shaft assembly, and jackshaft bracket).
3. Check belt and chain tension.
4. Check and tighten all nuts and bolts as necessary.
5. Check tire pressure and add air if needed.

To Start the Aerator

1. Disengage the belt tensioner.
2. Turn the engine switch to "On" and start the engine.
3. Adjust the throttle to a low rpm, firmly grip the handle, and engage the belt tensioner.
4. To shut off the machine, disengage the belt tensioner and turn the engine switch to "Off."

NOTES

Exercise 37 Activity

Student Name _____ Date _____ Score _____

Evaluation

Consideration	Points	Student Score	Instructor Score
Read owner's manual prior to operation	25		
Performed safety and engine inspection prior to using equipment	25		
Operated lawn aerator per instructions provided by teacher	25		
Followed safety precautions while using equipment	25		
Total	100		

NOTES

Operating a Walk-Behind Blower

OBJECTIVE

To safely operate a walk-behind blower.

SKILLS

After studying this unit, you should be able to:

- Operate a walk-behind blower in a proper and safe manner.

MATERIALS NEEDED

Walk-behind blower
Lawn practice area

INTRODUCTION

Blowers have become a useful piece of equipment for landscapers. They allow the use of sweeping leaves and debris from landscapes in a time-efficient manner. Blowers can be used to tidy up after mowing, blow lawn clippings off sidewalks, or clear leaf debris from lawns prior to mowing. With any type of landscape equipment, it is important to read the owner's or manufacturer's manual. Take time to understand the safety precautions you will need to use to operate the blower. Also, take time to understand how to effectively maintain your equipment to allow for longer use of the blower.

PUT IT INTO PRACTICE

CAUTION

Use extreme care when handling the walk-behind blower! It should be operated
by trained individuals only!

1. Read the following information.

2. Instructor will demonstrate how to properly and safely operate a walk-behind blower.

3. Practice the correct techniques for operating and maintaining the walk-behind blower as approved by the instructor. Please review the Evaluation rubric before beginning.

Safety Precautions

1. Read the manufacturer's manual thoroughly before operating or maintaining the blower.

2. Always wear safety glasses and long pants when operating or maintaining this machine. Do not wear loose-fitting clothing.

3. Always disconnect the spark plug wire before adjusting or maintaining this machine.

4. When operating, keep your hands and feet away from the side discharge and your hands and clothing away from the fan guard.

5. Do not remove any shields, guards, or decals. If a shield, guard, or decal is damaged or does not function, repair or replace it before operating the blower.

6. Altering this machine in any way may cause injury to the operator or bystanders.

7. Keep people and pets away from the blower when in use.

8. When operating, never direct the air stream at walls or vertical objects that can deflect debris back at you.

9. When operating, never direct the air stream at painted objects.

10. If a breakdown occurs, shut off the engine immediately, disconnect the spark plug wire, and do not connect until repairs have been made.

11. Never run the engine indoors without adequate ventilation. Exhaust fumes are deadly.

12. Stop the engine and allow it to cool before refilling the gas tank.

13. Gasoline is extremely dangerous and flammable. Do not permit open flames or sparks near the engine at any time.

14. Keep the blower and especially the engine air screen and cooling fins clean and free of grease, grass, and leaves to reduce the chance of fire and to permit proper cooling.

OPERATING INSTRUCTIONS

Prior to Each Day's Use

1. Disconnect the engine's spark plug wire.

2. Position the blower on a level surface and check the engine oil level. Clean the cooling air intake screen. Replenish the fuel.

3. Check and tighten all nuts and bolts as necessary.

4. Check tire pressure and add air if needed.

To Start the Blower

1. Connect the engine's spark plug wire.

2. Move the engine choke control down to "Choke." Move the throttle control about 30 degrees counterclockwise.

3. Grasp the starter grip, and pull the cord slowly until the starter engages and then pull rapidly to overcome compression, prevent kickback, and start the engine. Allow the cord to recoil slowly. Repeat if necessary.

4. Move the choke control up to the run position after starting and when the engine has warmed up, adjust the throttle to produce the amount of sweeping power required. At maximum throttle, the engine reaches the maximum speed of 3600 rpm, which produces a fan tip velocity of 175 mph and the maximum blowing capacity of 2500 cubic feet per minute (cfm).

5. Stand behind the blower, and move to the point where you want to begin sweeping. The sweeping action is equally effective whether the machine is pulled or pushed, but keep the front of the machine down and close to the work surface to blow out holes and crevices.

6. To shut off the machine, move the engine throttle control clockwise to the "Off" position.

7. After use, thoroughly clean the machine, especially the engine's air intake screen and cooling fins.

Engine Service Maintenance

Every 25 hours:	**Servicing Air Filter:**
	Service the foam air filter by washing in soap and water, and then rinse and dry. Re-oil with clean engine oil.
	Change the oil (check manufacturer's manual). Change the oil more frequently under dusty conditions.
Every 50 hours:	Service the fuel filter.
Every 100 hours:	**Servicing Air Filter:**
	Service the air filter cleaner paper element by gently tapping on the flat side. Do not wash or use pressurized air. Replace the paper element each year or more often under dusty conditions.

NOTES

Exercise 38 Activity

Student Name _____ Date _____ Score _____

Evaluation

Consideration	Points	Student Score	Instructor Score
Read owner's manual prior to operation	25		
Performed safety and engine inspection prior to starting blower	25		
Correctly and safely started and shut off the machine	25		
Performed a slow and effective sweeping motion while operating equipment per instructions provided by teacher	25		
Total	100		

NOTES

Lawn Renovation (Reseeding)

OBJECTIVE

To reseed a presently established lawn that is thinning, or fill in where dead spots have occurred.

SKILLS

After studying this unit, you should be able to:

- Reseed to improve appearance of established lawn.

MATERIALS NEEDED

Lawn renovator

Selected lawn seed for your area

Soil test

Lime or sulfur

Fertilizer 5-10-5

Lawn sprinkler and rain gauge

Introductory Horticulture, 9th Edition

Landscaping Principles & Practices, 8th Edition

INTRODUCTION

Reseeding an existing lawn is important to keep the lawn looking full and healthy. Bald spots or thin spots allow for weeds to grow and thrive. Once weeds enter into your lawn, you will have a difficult time controlling them, so it is important to look for bald and thinning spots and take care of them before weeds begin to sprout. Reseeding will allow you to patch these areas and create a fuller lawn.

PUT IT INTO PRACTICE

Practice the correct reseeding techniques as approved by the instructor. Please review the Evaluation rubric before beginning.

1. Soil-test the area to be renovated for pH and N, P, K.

2. Apply lime or sulfur to correct the pH to a steady 6.5.

3. Using a lawn renovator, reseed the lawn area.

4. Set up the sprinkler and rain gauge to apply 1 inch of water at each watering.

5. Observe for germination of the grass in about 8–10 days.

NOTES

Exercise 39 Activity

Student Name _____ Date _____ Score _____

Evaluation

Consideration	Points	Student Score	Instructor Score
Properly reseeded lawn area, using seeding technique demonstrated by instructor	50		
Properly set up irrigation to allow for germination of seed	50		
Total	100		

NOTES

Lawn Renovation (Overseeding)

OBJECTIVE

To overseed an established lawn or athletic field of thinning turf.

SKILLS

After studying this unit, you should be able to:

- Overseed to improve the appearance of an established lawn or thinning turf on athletic field.

MATERIALS NEEDED

Lawn aerator

Hi-wheeled cyclone spreader

Selected lawn seed for your geographical area

Soil-test kit or a commercial soil-test results recommendation sheet (e.g., CLC Labs)

Lime or sulfur (for pH correction)

Lawn starter fertilizer

Lawn sprinkler and rain gauge

Chain link fence section, 4 × 5–feet, for dragging

Introductory Horticulture, 9th Edition

Landscaping: Principles and Practices, 8th Edition

INTRODUCTION

Overseeding a large turf area allows for the opportunity to thicken areas where the turf may have thinned out. Overseeding in a large area requires the use of heavy equipment instead of small hand seeders. Overseeding should be completed before weeds begin to sprout. For turf to keep its beauty in high-traffic areas or for sports surfaces, it is important to make overseeding a priority in your maintenance schedule.

PUT IT INTO PRACTICE

Practice the correct overseeding techniques as approved by the instructor. Please review the Evaluation rubric before beginning.

1. Soil-test the area to be overseeded for pH and N, P, K.

2. Correct soil pH and N, P, K to the soil-test recommendation.

3. Using the lawn aerator, aerate north–south and east–west. This will ensure good penetration.

4. Using a Lesco high wheel spreader, sow the grass seed over the area, using local recommendations on seed and seeding rate.

5. Using a 4 × 5–foot section of a chain link fence, drag the seeded area to ensure good soil–seed contact.

6. Set up the sprinkler and rain gauge to apply 1 inch of water over the entire seeded area.

7. Observe for germination of the seed in 8–10 days.

NOTES

Exercise 40 Activity

Student Name _____ Date _____ Score _____

Evaluation

Consideration	Points	Student Score	Instructor Score
Properly overseeded lawn area using seeding technique demonstrated by instructor	50		
Properly set up irrigation to allow for germination of seed	50		
Total	100		

NOTES